Springer Proceedings in Physics

Volume 164

More information about this series at http://www.springer.com/series/361

E.K. Polychroniadis · Ahmet Yavuz Oral
Mehmet Ozer
Editors

2nd International Multidisciplinary Microscopy and Microanalysis Congress

Proceedings of InterM, October 16–19, 2014

Springer

Editors
E.K. Polychroniadis
Department of Physics
Aristotle University of Thessaloniki
Thessaloniki
Greece

Mehmet Ozer
Department of Physics
Istanbul Kultur University
Istanbul, Bakirkoy
Turkey

Ahmet Yavuz Oral
Department of Materials Science and Engineering
Gebze Technical University
Gebze, Kocaeli
Turkey

ISSN 0930-8989 ISSN 1867-4941 (electronic)
Springer Proceedings in Physics
ISBN 978-3-319-16918-7 ISBN 978-3-319-16919-4 (eBook)
DOI 10.1007/978-3-319-16919-4

Library of Congress Control Number: 2015936290

Springer Cham Heidelberg New York Dordrecht London

Printed on acid-free paper

Springer International Publishing AG Switzerland is part of Springer Science+Business Media (www.springer.com)

Preface

The second International Multidisciplinary Microscopy and Microanalysis Congress & Exhibition (INTERM 2014) provided all these scientists the opportunity to meet, present their work, discuss and mutually interact in order to enhance and promote their research work.

This volume, published by Springer, includes all the papers presented at this Congress, held in Liberty Hotels Lykia Oludeniz, Turkey, October 16–19, 2014.

On behalf of the organizing committee we would like to thank all the plenary and invited speakers for their valuable contribution and especially Professor Gustaaf Van Tendeloo (EMAT, University of Antwerp, Belgium) for his excellent opening lecture.

We would also like to thank TURA Tourism for their support in the organization of the Congress, as well as our sponsors: JEOL Ltd., NanoMEGAS, and PVA TePla. Finally, we would like to thank the publishers for the quality of this edition.

Thessaloniki, Greece E.K. Polychroniadis
Gebze, Turkey Ahmet Yavuz Oral
Istanbul, Turkey Mehmet Ozer

Organization

International Scientific Committee

Prof. Ernesto Diéguez Delgado, Universidad Autónoma de Madrid, Spain
Prof. Wolfgang Eberhardt, Technische Universität Berlin, Germany
Dr. Konstantinos, P. Giannakopoulos, National Centre for Scientific Research "Demokritos", Greece
Prof. Tomasz Goryczka, University of Silesia, Poland
Prof. Christos B. Lioutas, Aristotle University of Thessaloniki, Greece
Dr. Stavros Nicolopoulos, NanoMEGAS, Belgium
Prof. Emer. E.K. Polychroniadis, Aristotle University of Thessaloniki, Greece
Assoc. Prof. Dr. Ahmet Yavuz Oral, Gebze Technical University, Turkey
Prof. Danuta Stróż, University of Silesia, Poland
Prof. Gustaaf Van Tendeloo, EMAT, University of Antwerp, Belgium

Organizing Committee

E.K. Polychroniadis, Aristotle University of Thessaloniki, Greece
Ahmet Yavuz Oral, Gebze Technical University, Turkey
Mehmet Ozer, Istanbul Kültür University (Scientific Secretariat), Turkey
Z. Banu Bahsi, Gebze Institute of Technology, Turkey
Ersin Kayahan, Kocaeli University, Turkey

Sponsorship & Exhibitation

The Company—JEOL Ltd.

The company NanoMEGAS

The company PVA TePla

Conference Organizing Company

TURA TOURISM LTD.

Cumhuriyet Cad. No: 109/A Elmadag—Sisli/Istanbul
E-mail: interem2013@turaturizm.com.tr
Phone: +90 212 241 27 00
Fax: +90 212 241 29 89
Web: www.turaturizm.com.tr

Contents

Part I Applications of Microscopy in the Physical Sciences

Electron Microscopy Study of Thermoelectric $(Bi_xSb_{1-x})_2Te_3$ Thin Film 3
Aikaterini Breza, Christos B. Lioutas and John Giapintzakis

Structural Characterization of Layers for Advanced Non-volatile Memories 9
K. Giannakopoulos, J. Giannopoulos, P. Bousoulas, E. Verrelli and D. Tsoukalas

Advanced Technology for Analytical Electron Microscopy by Using Aberration Corrected Transmission Electron 19
Guillaume Brunetti

TEM Characterization of a Complex Twinning System in 3C–SiC 25
Mamour Sall, Narendraraj Chandran, Anastasia Terzidou, Christos B. Lioutas and E.K. Polychroniadis

Microstructural Evaluation of Suspension Thermally Sprayed WC-Co Nanocomposite Coatings 31
R. Ahmed, N.H. Faisal, O. Ali, Nayef M. Al-Anazi, S. Al-Mutairi, S. Mamour, E.K. Polychroniadis and M.F.A. Goosen

Wavelength Depended Speckle Correlation Analyses of Engineered Metal Surfaces 39
Ersin Kayahan

Effect of V_2O_5 Additives to the Sintering of Y_2O_3 51
Tarik Talib Issa, Kawakib Jassim Majeed and Jenan M. Hasan

Integrating Microscopic Analysis into Existing Quality Assurance Processes 57
Peter Frühberger, Thomas Stephan and Jürgen Beyerer

Effect of Annealing Temperature on the Structural and Magnetic Properties of Terbium Iron Garnet Thin Films Prepared by Sol-Gel Method 65
Ftema W. Aldbea and N.B. Ibrahim

Fibrous Growth of Chloride Minerals on Diatomite Saturated with a Brine 73
Tomasz Toboła, Marek Rembiś, Beata Figarska-Warchoł and Grażyna Stańczak

Fractal Characteristics of the Pore Network in Diatomites Using Mercury Porosimetry and Image Analysis 79
Grażyna Stańczak, Marek Rembiś, Beata Figarska-Warchoł and Tomasz Toboła

Development of an Off-Axis Digital Holographic Microscope for Large Scale Measurement in Fluid Mechanics 91
K.F. Tamrin, B. Rahmatullah and S.M. Samuri

SEM-EDS Observation of Structure Changes in Synthetic Zeolites Modified for CO_2 Capture Needs 97
Magdalena Wdowin, Rafal Panek and Wojciech Franus

SEM Investigation of Microstructures in Hydration Products of Portland Cement 105
Wojciech Franus, Rafal Panek and Magdalena Wdowin

Some Properties of 1.3343 Steel Treated by Pulse Plasma Technique 113
Tuğçe Gökkaya and Yıldız Yaralı Özbek

Investigation of Microstructure of Ceramics Produced from Gabbro and Zeolite Raw Materials 121
Nuray Canikoğlu and A. Şükran Demirkıran

The Effect of MgO and MgO–Al_2O_3 on Zirconia Produced by Precipitation Method 129
M. Ipek

Microstructural Investigation of IF Steels Joined by Metal Inert Gas Brazing 137
Mehmet Ekici, Faruk Varol, Uğur Özsaraç and Salim Aslanlar

An ESEM/EDX Methodology for the Study of Additive Adsorption at the Polymer-Air Interface 145
Omar Islam, Clive R. Siviour and Kalin I. Dragnevski

Enhanced Confocal Fluorescence Microscope Performance Using a differential pinhole 153
Rohan Kakade, John G. Walker and Andrew J. Phillips

Thermal Neutron Detection by Entrapping ^{6}LiF Nanodiamonds in Siloxane Scintillators 161
M. Degerlier, S. Carturan, T. Marchi, M. Dalla Palma, F. Gramegna, G. Maggioni, M. Cinausero and A. Quaranta

High Temperature Reliability of Ta-Based and TiW-Based Diffusion Barriers 169
Nando Budhiman, Ulrich Schürmann, Björn Jensen, Steffen Chemnitz, Lorenz Kienle and Bernhard Wagner

The Direct Observation of Grain Refinement Mechanism in Advanced Multicomponent γ-TiAl Based Structural Intermetallics Doped with Boron 175
A.V. Kartavykh, M.V. Gorshenkov and D.A. Podgorny

Low Temperature Resistivity of the Rare Earth Diborides (Er, Ho, Tm)B_2 183
Jumat B. Kargin, C.R. Sebastian Haines, Matthew J. Coak, Cheng Liu, Alexander V. Matovnikov, Vladimir V. Novikov, Alexander N. Vasiliev and Siddharth S. Saxena

Influence of Mechanical Parameters on the Friction and Wear of Sliding Brass-Steel Couple 187
C. Boubechou, A. Bouchoucha and H. Zaidi

Measuring the Degree of Sensitization (DOS) Using an Electrochemical Technique 197
Mokhtar B. Abuzriba and Salem M. Musa

Substitution for Chromium and Nickel in Austenitic Stainless Steels 205
Mokhtar B. Abuzriba and Salem M. Musa

Part II Applications of Microscopy in the Biological Sciences

Ca, P and Collagen Fibrils Period Measurements in the Vertebras of Lordotic *Sparus aurata* 217
Panagiotis Berillis and Nikolaos Panagiotopoulos

Use of Bone Marrow-Derived Mesenchymal Stem Cells in Improving Thioacetamide Induced Liver Fibrosis in Rats 223
Fatma A.A. Mansour, Iman Shaheed and Nabiha R.A. Hassan

Electrochemical Detection of Nicotine Using Cerium Nanoparticles Modified Carbon Paste Sensor and Anionic Surfactant 229
A.M. Fekry, S.M. Azab, M. Shehata and M.A. Ameer

Exploring the Antibiotic Effects in Bacterial Biofilms by Epifluorescence and Scanning Electron Microscopy 241
Luciana Calheiros Gomes, Laura Nunes Silva, Manuel Simões, Luís Ferreira de Melo and Filipe José Mergulhão

Quantitative Confocal Microscopy Analysis as a Basis for Search and Study of Potassium Kv1.x Channel Blockers 249
Alexey V. Feofanov, Kseniya S. Kudryashova, Oksana V. Nekrasova, Alexander A. Vassilevski, Alexey I. Kuzmenkov, Yuliya V. Korolkova, Eugene V. Grishin and Mikhail P. Kirpichnikov

Analysis of Nucleosome Transcription Using Single-Particle FRET . . . 255
Alexey V. Feofanov, Kseniya S. Kudryashova, Oleg V. Chertkov, Dmitry V. Nikitin, Nikolai A. Pestov, Olga I. Kulaeva, Vasily M. Studitsky and Mikhail P. Kirpichnikov

Index 261

Contributors

Mokhtar B. Abuzriba Department of Materials and Metallurgical Engineering, University of Tripoli, Tripoli, Libya

R. Ahmed College of Engineering, Alfaisal University, Riyadh, Saudi Arabia; School of Engineering and Physical Sciences, Heriot-Watt University, Edinburgh, UK

Nayef M. Al-Anazi Materials Performance Unit, Research & Development Centre, Dhahran, Saudi Arabia

S. Al-Mutairi Materials Performance Unit, Research & Development Centre, Dhahran, Saudi Arabia

Ftema W. Aldbea Faculty of Science and Technology, Universiti Kebangsaan Malaysia, Bangi, Malaysia; Faculty of Science, Sebha University, Sebha, Libya

O. Ali School of Engineering and Physical Sciences, Heriot-Watt University, Edinburgh, UK

M.A. Ameer Chemistry Department, Faculty of Science, Cairo University, Giza, Egypt

Salim Aslanlar Department of Metallurgical and Materials Engineering, Technology Faculty, Sakarya University, Sakarya, Turkey

S.M. Azab Pharmaceutical Chemistry Department, National Organization for Drug Control and Research (NODCAR), Giza-29, Egypt

Panagiotis Berillis Department of Ichthyology and Aquatic Environment, School of Agricultural Sciences, University of Thessaly, Volos, Greece

Jürgen Beyerer Fraunhofer Institute of Optronics, System Technologies and Image Exploitation IOSB, Karlsruhe, Germany

C. Boubechou Faculté de Technologie, Département de Génie Mécanique, Université de 20 Août 1955, Skikda, Algeria

A. Bouchoucha Laboratoire de Mécanique, Faculté des Sciences de la Technologie, Département de Mécanique Engineering, Université Constantine 1, Constantine, Algeria

P. Bousoulas Department of Applied Physics, National Technical University of Athens, Athens, Greece

Aikaterini Breza Department of Physics, Aristotle University of Thessaloniki, Thessaloniki, Greece

Guillaume Brunetti JEOL (Europe) SAS—Espace Claude Monet, Croissy-sur-Seine, France

Nando Budhiman Institute of Material Science, Christian-Albrechts-Universität zu Kiel, Kiel, Germany

Nuray Canikoğlu Department of Metallurgy and Materials Engineering, Engineering Faculty, Sakarya University, Sakarya, Turkey

S. Carturan INFN—Laboratori Nazionali di Legnaro, Legnaro, Italy; Department of Physics and Astronomy, University of Padova, Padua, Italy

Narendraraj Chandran Department of Physics, Aristotle University of Thessaloniki, Thessaloniki, Greece

Steffen Chemnitz Institute of Material Science, Christian-Albrechts-Universität zu Kiel, Kiel, Germany; Fraunhofer Institute for Silicon Technology (ISiT), Itzehoe, Germany

Oleg V. Chertkov Biological Faculty, Lomonosov Moscow State University, Moscow, Russia; Shemyakin-Ovchinnikov Institute of Bioorganic Chemistry, Russian Academy of Sciences, Moscow, Russia

M. Cinausero INFN—Laboratori Nazionali di Legnaro, Legnaro, Italy

Matthew J. Coak Cavendish Laboratory, University of Cambridge, Cambridge, UK

M. Dalla Palma INFN—Laboratori Nazionali di Legnaro, Legnaro, Italy; Department of Industrial Engineering, University of Trento, Trento, Italy

M. Degerlier Science and Art Faculty Physics Department, Nevsehir Haci Bektas Veli University, Nevsehir, Turkey

A. Şükran Demirkıran Department of Metallurgy and Materials Engineering, Engineering Faculty, Sakarya University, Sakarya, Turkey

Kalin I. Dragnevski Laboratory for In-situ Microscopy and Analysis, Department of Engineering Science, University of Oxford, Oxford, UK

Mehmet Ekici Vocational School of Yalova, Yalova University, Yalova, Turkey

N.H. Faisal College of Engineering, Alfaisal University, Riyadh, Saudi Arabia; School of Engineering, Robert Gordon University, Aberdeen, UK

A.M. Fekry Chemistry Department, Faculty of Science, Cairo University, Giza, Egypt

Alexey V. Feofanov Shemyakin-Ovchinnikov Institute of Bioorganic Chemistry, Russian Academy of Sciences, Moscow, Russia; Biological Faculty, Lomonosov Moscow State University, Moscow, Russia

Beata Figarska-Warchoł Faculty of Geology, Geophysics and Environmental Protection, AGH University of Science and Technology, Kraków, Polska

Wojciech Franus Division of Geotechnics, Lublin University of Technology, Lublin, Poland

Peter Frühberger Fraunhofer Institute of Optronics, System Technologies and Image Exploitation IOSB, Karlsruhe, Germany

K. Giannakopoulos Institute of Nanoscience and Nanotechnology, NCSR "Demokritos", Athens, Greece

J. Giannopoulos Institute of Nanoscience and Nanotechnology, NCSR "Demokritos", Athens, Greece; Department of Applied Physics, National Technical University of Athens, Athens, Greece

John Giapintzakis Department of Mechanical Engineering, University of Cyprus, Nicosia, Cyprus

Luciana Calheiros Gomes LEPABE, Department of Chemical Engineering, Faculty of Engineering, University of Porto, Porto, Portugal

M.F.A. Goosen Office of Research & Graduate Studies, Alfaisal University, Riyadh, Saudi Arabia

M.V. Gorshenkov National University of Science and Technology "MISIS", Moscow, Russia

F. Gramegna INFN—Laboratori Nazionali di Legnaro, Legnaro, Italy

Eugene V. Grishin Shemyakin-Ovchinnikov Institute of Bioorganic Chemistry, Russian Academy of Sciences, Moscow, Russia

Tuğçe Gökkaya Department of Metallurgy and Materials Engineering, Esentepe Campus, Sakarya University, Sakarya, Turkey

C.R. Sebastian Haines Cavendish Laboratory, University of Cambridge, Cambridge, UK

Jenan M. Hasan Physics Department, College of Science, University of Baghdad, Jadiriya, Baghdad, Iraq

Nabiha R.A. Hassan Department of Pathology, Faculty of Veterinary Medicine, Cairo University, Cairo, Egypt

N.B. Ibrahim Faculty of Science and Technology, Universiti Kebangsaan Malaysia, Bangi, Malaysia

M. Ipek Department of Metallurgy and Materials Engineering, Engineering Faculty, Esentepe Campus, Sakarya University, Sakarya, Turkey

Omar Islam Laboratory for In-situ Microscopy & Analysis, Department of Engineering Science, University of Oxford, Oxford, UK

Tarik Talib Issa Physics Department, College of Science, University of Baghdad, Jadiriya, Baghdad, Iraq

Björn Jensen Fraunhofer Institute for Silicon Technology (ISiT), Itzehoe, Germany

Rohan Kakade Applied Optics Group, Faculty of Engineering, The University of Nottingham, University Park, Nottingham, UK

Jumat B. Kargin Cavendish Laboratory, University of Cambridge, Cambridge, UK; Eurasian National University, Astana, Kazakhstan

A.V. Kartavykh National University of Science and Technology "MISIS", Moscow, Russia

Ersin Kayahan Electro-Optic and System Engineering, Kocaeli University, Umuttepe, Kocaeli, Turkey; Kocaeli University-LATARUM Laboratory, Yenikoy, Kocaeli, Turkey; Barbaros Denizcilik YO, Kocaeli University, Karamürsel, Kocaeli, Turkey

Lorenz Kienle Institute of Material Science, Christian-Albrechts-Universität zu Kiel, Kiel, Germany

Mikhail P. Kirpichnikov Shemyakin-Ovchinnikov Institute of Bioorganic Chemistry, Russian Academy of Sciences, Moscow, Russia; Biological Faculty, Lomonosov Moscow State University, Moscow, Russia

Yuliya V. Korolkova Shemyakin-Ovchinnikov Institute of Bioorganic Chemistry, Russian Academy of Sciences, Moscow, Russia

Kseniya S. Kudryashova Shemyakin-Ovchinnikov Institute of Bioorganic Chemistry, Russian Academy of Sciences, Moscow, Russia; Biological Faculty, Lomonosov Moscow State University, Moscow, Russia

Olga I. Kulaeva Biological Faculty, Lomonosov Moscow State University, Moscow, Russia; Cancer Epigenetics Program, Fox Chase Cancer Center, Philadelphia, USA

Alexey I. Kuzmenkov Shemyakin-Ovchinnikov Institute of Bioorganic Chemistry, Russian Academy of Sciences, Moscow, Russia

Christos B. Lioutas Department of Physics, Aristotle University of Thessaloniki, Thessaloniki, Greece

Cheng Liu Cavendish Laboratory, University of Cambridge, Cambridge, UK

G. Maggioni INFN—Laboratori Nazionali di Legnaro, Legnaro, Italy; Department of Physics and Astronomy, University of Padova, Padua, Italy

Kawakib Jassim Majeed Physics Department, College of Science, University of Baghdad, Jadiriya, Baghdad, Iraq

S. Mamour Department of Physics, Aristotle University of Thessaloniki, Thessaloniki, Greece

Fatma A.A. Mansour Department of Pathology, Faculty of Veterinary Medicine, Cairo University, Cairo, Egypt

T. Marchi INFN—Laboratori Nazionali di Legnaro, Legnaro, Italy

Alexander V. Matovnikov Bryansk State University, Bryansk, Russia

Luís Ferreira de Melo LEPABE, Department of Chemical Engineering, Faculty of Engineering, University of Porto, Porto, Portugal

Filipe José Mergulhão LEPABE, Department of Chemical Engineering, Faculty of Engineering, University of Porto, Porto, Portugal

Salem M. Musa Department of Materials and Metallurgical Engineering, University of Tripoli, Tripoli, Libya

Oksana V. Nekrasova Shemyakin-Ovchinnikov Institute of Bioorganic Chemistry, Russian Academy of Sciences, Moscow, Russia

Dmitry V. Nikitin Biological Faculty, Lomonosov Moscow State University, Moscow, Russia

Vladimir V. Novikov Bryansk State University, Bryansk, Russia

Nikolaos Panagiotopoulos Department of Physics, University of Ioannina, Ioannina, Greece

Rafal Panek Division of Geotechnics, Lublin University of Technology, Lublin, Poland

Nikolai A. Pestov Biological Faculty, Lomonosov Moscow State University, Moscow, Russia

Andrew J. Phillips Applied Optics Group, Faculty of Engineering, The University of Nottingham, University Park, Nottingham, UK

D.A. Podgorny National University of Science and Technology "MISIS", Moscow, Russia

E.K. Polychroniadis Department of Physics, Aristotle University of Thessaloniki, Thessaloniki, Greece

A. Quaranta INFN—Laboratori Nazionali di Legnaro, Legnaro, Italy; Department of Industrial Engineering, University of Trento, Trento, Italy

B. Rahmatullah Computing Department, Faculty of Arts, Computing and Creative Industry, Universiti Pendidikan Sultan Idris, Perak, Malaysia

Marek Rembiś Faculty of Geology, Geophysics and Environmental Protection, AGH University of Science and Technology, Kraków, Polska

Marek Rembiś Faculty of Geology, Geophysics and Environmental Protection, AGH University of Science and Technology, Krakow, Poland

Mamour Sall Department of Physics, Aristotle University of Thessaloniki, Thessaloniki, Greece

S.M. Samuri Computing Department, Faculty of Arts, Computing and Creative Industry, Universiti Pendidikan Sultan Idris, Perak, Malaysia

Siddharth S. Saxena Cavendish Laboratory, University of Cambridge, Cambridge, UK

Ulrich Schürmann Institute of Material Science, Christian-Albrechts-Universität zu Kiel, Kiel, Germany

Iman Shaheed Department of Pathology, Faculty of Veterinary Medicine, Cairo University, Cairo, Egypt

M. Shehata Chemistry Department, Faculty of Science, Cairo University, Giza, Egypt

Laura Nunes Silva LEPABE, Department of Chemical Engineering, Faculty of Engineering, University of Porto, Porto, Portugal

Manuel Simões LEPABE, Department of Chemical Engineering, Faculty of Engineering, University of Porto, Porto, Portugal

Clive R. Siviour Laboratory for In-situ Microscopy & Analysis, Department of Engineering Science, University of Oxford, Oxford, UK

Grażyna Stańczak Faculty of Geology, Geophysics and Environmental Protection, AGH University of Science and Technology, Kraków, Polska

Thomas Stephan Fraunhofer Institute of Optronics, System Technologies and Image Exploitation IOSB, Karlsruhe, Germany

Vasily M. Studitsky Biological Faculty, Lomonosov Moscow State University, Moscow, Russia; Cancer Epigenetics Program, Fox Chase Cancer Center, Philadelphia, USA

K.F. Tamrin Computing Department, Faculty of Arts, Computing and Creative Industry, Universiti Pendidikan Sultan Idris, Perak, Malaysia

Anastasia Terzidou Department of Physics, Aristotle University of Thessaloniki, Thessaloniki, Greece

Tomasz Toboła Faculty of Geology, Geophysics and Environmental Protection, AGH University of Science and Technology, Krakow, Poland

D. Tsoukalas Department of Applied Physics, National Technical University of Athens, Athens, Greece

Faruk Varol Vocational School of Karasu, Sakarya University, Sakarya, Turkey

Alexander N. Vasiliev Moscow State University, Moscow, Russia

Alexander A. Vassilevski Shemyakin-Ovchinnikov Institute of Bioorganic Chemistry, Russian Academy of Sciences, Moscow, Russia

E. Verrelli Department of Applied Physics, National Technical University of Athens, Athens, Greece

Bernhard Wagner Institute of Material Science, Christian-Albrechts-Universität zu Kiel, Kiel, Germany; Fraunhofer Institute for Silicon Technology (ISiT), Itzehoe, Germany

John G. Walker Applied Optics Group, Faculty of Engineering, The University of Nottingham, University Park, Nottingham, UK

Magdalena Wdowin The Mineral and Energy Economy Research Institute of the Polish Academy of Sciences, Kraków, Poland

Yıldız Yaralı Özbek Department of Metallurgy and Materials Engineering, Esentepe Campus, Sakarya University, Sakarya, Turkey

H. Zaidi Laboratoire LMS (UMR-6610-CNRS), SP2MI, Téléport 2, Boulevard Marie et Pierre Curie, Université of Poitiers, Futuroscope Chasseneuil Cedex, Poitiers, France

Uğur Özsaraç Technology Faculty, Department of Metallurgical and Materials Engineering, Sakarya University, Sakarya, Turkey

Part I
Applications of Microscopy in the Physical Sciences

Electron Microscopy Study of Thermoelectric $(Bi_xSb_{1-x})_2Te_3$ Thin Film

Aikaterini Breza, Christos B. Lioutas and John Giapintzakis

Abstract In the present work, transmission electron microscopy (TEM) techniques were used in order to study the morphology and investigate the structural properties of a $(Bi_xSb_{1-x})_2Te_3$ (BST) thin film. The sample was fabricated on silicon substrate by pulsed laser deposition (PLD) method. Towards this aim, cross sectional and planar view samples were prepared, suitable for High Resolution Electron Microscopy. Results revealed a polycrystalline, c-axis oriented film with coherent boundaries between neighboring columnar grains.

1 Introduction

In the past decades, thermoelectric films have received increasing interests due to their application on electronic devices in the micro- heating and cooling areas [1]. Bismuth telluride based thin films have gained growing interest since they have attractive thermoelectric properties and can potentially contribute to saving energy issues [2]. Generally, the performance of thermoelectric materials is characterized by the dimensionless figure-of-merit, ZT, defined as ZT = S2σT/κ, where S is the Seebeck coefficient, σ is the electrical conductivity, T is the absolute temperature and κ is the total thermal conductivity with contributions from the lattice thermal conductivity (κL) and the electronic thermal conductivity (κe). Theoretical

A. Breza · C.B. Lioutas (✉)
Department of Physics, Aristotle University of Thessaloniki, 54124 Thessaloniki, Greece
e-mail: lioutas@physics.auth.gr

A. Breza
e-mail: abreza@physics.auth.gr

J. Giapintzakis
Department of Mechanical Engineering, University of Cyprus, 75 Kallipoleos Avenue, PO Box 20537, 1678 Nicosia, Cyprus
e-mail: giapintz@ucy.ac.cy

E.K. Polychroniadis et al. (eds.), *2nd International Multidisciplinary Microscopy and Microanalysis Congress*, Springer Proceedings in Physics 164,
DOI 10.1007/978-3-319-16919-4_1

approach has shown that nanostructuring Bi_2Te_3 alloys enhances their performance compared to the corresponding bulk materials, as quantum confinement effect leads in reduction in thermal conductivity [3]. In this direction, researcher groups have fabricated several types of low-dimension structures and confirmed that spatial confinement significantly improves ZT [4–9].

From a structural point of view, bismuth antimony telluride has the same rhombohedral tetradymite structure as Bi_2Te_3 of the space group R-3m. It could be described as a hexagonal unit cell consisting of five covalently bonded atomic lamellae stacked in the rhombohedral [111] direction [10]. It is worth noting that nanostructuring these materials is due this anisotropic character. In this work, we report on the morphological characteristics and structural properties of a BST thin film.

2 Experimental Details

The BST thin film that we studied in this work, was deposited on n-type Si (1 0 0) single crystal substrate by PLD method. UV-excimer laser pulses with wavelength $\lambda = 248$ nm, pulse duration $\tau_L = 25$ ns, pulse repetition rate of 10 Hz and fluence $\Phi = 2$ J cm^{-2} were employed for target ablation. For depositions 700 laser pulses were used. The substrate was heated to 350 °C. The BST target was a polycrystalline pellet with density higher than 90 % of the theoretical density. The structural features of the film were studied by TEM using a JEOL 2011 transmission electron microscope, operating at 200 kV and having a point resolution of 1.94 Å. Specimens suitable for cross-sectional and planar view TEM observations, were prepared using well-known techniques, including mechanical griding followed by ion milling with Ar ions of energy of 4 keV (Gatan PIPS).

3 Results and Discussion

The main features of the film are shown on a typical cross-sectional picture of the film presented in Fig. 1a. As it can be seen from the lower part of the film, it consists of almost columnar crystals, growing from the substrate up to the surface of the sample, of sizes from 80 to 150 nm, while its thickness is measured at 330 nm. A closer look at the interface with the Si substrate reveals the presence of an amorphous SiO_2 layer, having thickness of about 6.5 nm. Study of the corresponding electron diffraction pattern (Fig. 1b) confirms the epitaxial grown of the film, which is c-axis of all crystals parallel to [001]Si direction. Furthermore, all diffraction spots derived from the film were identified to belong to $(Bi_{0.5}Sb_{0.5})_2Te_3$ compound (72-1835#PDF).

High Resolution TEM study was performed in order to clarify the structure of the interfaces between the nano-columns. A typical micrograph of a boundary

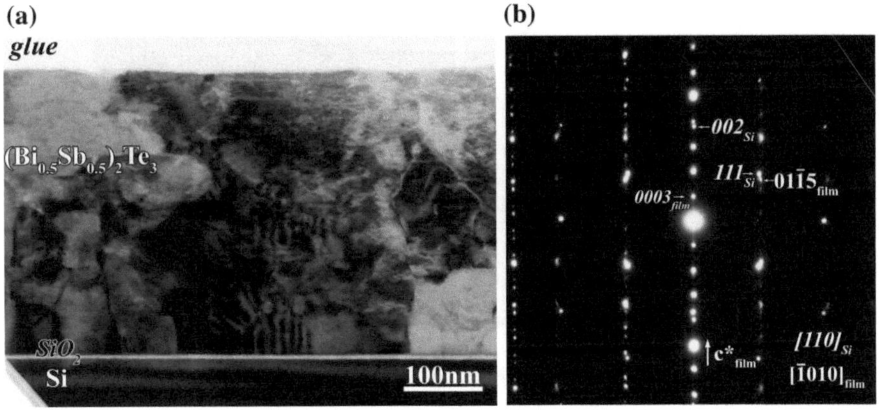

Fig. 1 **a** Bright-field image of film and substrate and **b** the corresponding electron diffraction pattern

Fig. 2 **a** A typical coherent boundary is shown, while **b** and **c** FFTs correspond to crystallites I and II respectively and reveal the orientation relation between the columns

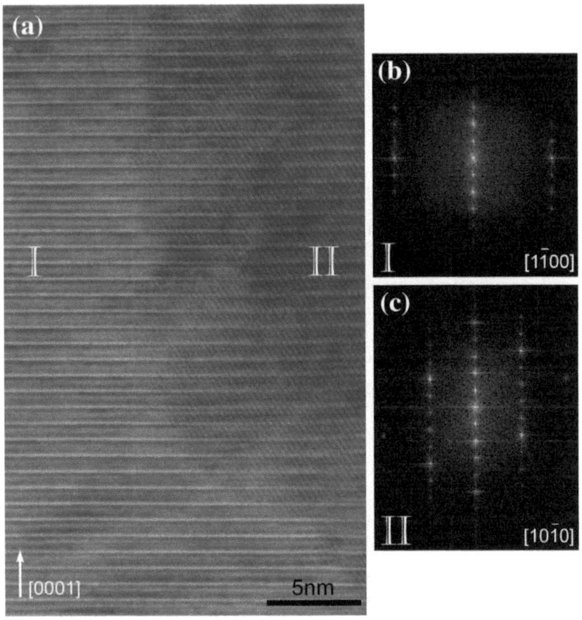

between two columns is given in Fig. 2a. It is worth noting that (0001) planes are coherent throughout the boundary. As identified by the corresponding FFTs (Fig. 2c, d), crystallites I and II are observed along [1-100] and [10-10] zones, respectively.

A planar-view image of the film is given in Fig. 3a, revealing the polygon shape of the crystals, observed along [0001] direction. In the corresponding electron diffraction pattern (Fig. 3b), diffraction rings confirm that it is polycrystalline.

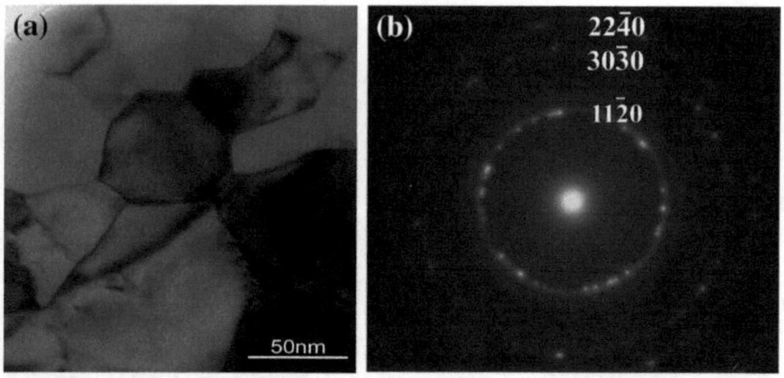

Fig. 3 **a** Bright-field image of the multigrains **b** the corresponding SAED of film

At the first view there is no preferential orientation of a-axis of the crystallites, regarding the substrate. Although, additional measurements of the angles between adjacent crystals, shows a tendency to form angles in good agreement with Coincide Site Lattice (CSL) theory.

4 Conclusion

We investigated by means of TEM the characteristics of a thermoelectric $(Bi_xSb_{1-x})_2Te_3$ thin film grown by pulsed laser deposition technique. Throughout the examined areas, the hexagonal $(Bi_{0.5}Sb_{0.5})_2Te_3$ compound (72-1835#PDF) was found. TEM micrographs and EDs showed that the 330 nm-thick film is poly-crystalline and consists of columnar crystals of 80–150 nm width with orientations close to that predicted from CLS theory. HRTEM images revealed the epitaxial relationship between film and substrate, which is $[0001]_{film}//[001]_{sub}$.

References

1. Fr. J. DiSalvo, Thermoelectric cooling and power generation, Science, **285**, 703–706 (1999)
2. C. Wood, Materials for thermoelectric energy conversion. Rep. Prog. Phys. **51**, 459–539 (1988)
3. L.D. Hicks, M.S. Dresselhaus, Effect of quantum-well structures on the thermoelectric figure of merit. Phys. Rev. B **47**, 12727–12731 (1993)
4. Y. Lan, B. Poudel, Y. Ma, D. Wang, M.S. Dresselhaus, G. Chen, Z. Ren, Structure study of bulk nanograined thermoelectric bismuth antimony telluride. Nano Lett. **9**, 1419–1422 (2009)
5. M. Takashiri, S. Tanaka, K. Miyazaki, Improved thermoelectric performance of highly-oriented nanocrystalline bismuth antimony telluride thin films. Thin Solid Films **519**, 619–624 (2010)

6. A. Purkayastha, F. Lupo, S. Kim, T. Borca-Tasciuc, G. Ramanath, Low-temperature, template-free synthesis of single-crystal bismuth telluride nanorods. Adv. Mater. **18**, 496–500 (2006)
7. M. Takashiri, S. Tanaka, K. Miyazaki, Growth of single-crystalline bismuth antimony telluride nanoplates on the surface of nanoparticle thin films. J. Cryst. Growth **372**, 199–204 (2013)
8. Y. Deng, Y. Xiang, Y. Song, Template-free Synthesis and transport properties of Bi_2Te_3 ordered nanowire arrays via a physical vapor process. Cryst. Growth Des. **9**, 3079–3082 (2009)
9. J.L. Mi, N. Lock, T. Sun, M. Christensen, M. Sondergaard, P. Hald, H.H. Hng, J. Ma, B.B. Iversen, Biomolecule-assisted hydrothermal synthesis and self-assembly of Bi_2Te_3 nanostring-cluster hierarchical structure. ACS Nano **4**, 2523–2530 (2010)
10. Y. Feutelais, B. Legendre, N. Rodier, V. Agafonov, A study of the phases in the bismuth tellurium system. Mater. Res. Bull. **28**, 591–596 (1993)

Structural Characterization of Layers for Advanced Non-volatile Memories

K. Giannakopoulos, J. Giannopoulos, P. Bousoulas,
E. Verrelli and D. Tsoukalas

Abstract Non-volatile memory cells are the devices with the most aggressive scaling on the market. For this reason the accurate characterization of their layer stacks is of great importance. We present a review of our recent work on a large variety of such stacks, for charge-trap and resistive memories, which have been characterized structurally with Transmission Electron Microscopy and Conducting Atomic Force Microscopy; we discuss the features of their structure on their function as memory elements.

1 Introduction

The technology for creating advanced memory cells plays a central role in the miniaturisation of electronic devices; new process nodes usually enter production with the manufacturing of memory cells. At this moment the 16 nm memory cells are in production. The recent slowdown of the transistor scaling has partially been compensated by the improvements in memory performance and by their integration on various chips. There is a fierce competition between memory manufacturers which is governed by their ability to introduce new processes on time that are improving the speed, the reliability and the density of the memory cells, while keeping the production and development cost at a low level.

Non-volatile memories are at the centre of this technology and during the last years Solid State Disks have become mainstream computer components. Flash memories have dominated this vast field for a long time, and because their scaling

K. Giannakopoulos (✉) · J. Giannopoulos
Institute of Nanoscience and Nanotechnology, NCSR "Demokritos", Aghia Paraskevi, 15310 Athens, Greece
e-mail: K.giannakopoulos@inn.demokritos.gr

J. Giannopoulos · P. Bousoulas · E. Verrelli · D. Tsoukalas
Department of Applied Physics, National Technical University of Athens, Iroon Polytechniou 9 Zografou, 15780 Athens, Greece

E.K. Polychroniadis et al. (eds.), *2nd International Multidisciplinary Microscopy and Microanalysis Congress*, Springer Proceedings in Physics 164,
DOI 10.1007/978-3-319-16919-4_2

has also been slowed down, the interest of the scientific community is shifting increasingly towards other technologies, such as the Resistive Memories. The resistive memory is currently the most promising non-volatile memory technology, as it offers low-power, reliable and denser memory cells with the option to exploit not only single level cell storage (storing one bit of information) but also multilevel cells, a feature that multiplies the effective information density storage.

From a Materials Science point of view, advanced non-volatile memories consist of nanometer scale stacks that can be processed with a large number of methods, in order to achieve the best balance of their electrical properties. Insulating, semiconducting and metallic properties are pushed to their limits while there is a constant search for new materials and device geometries. Of course, the introduction of new materials and process is not easily welcomed by the manufacturers, as it may lower the production yield and therefore increase significantly the cost of the devices. They prefer to work with "CMOS compatible" materials and processes, i.e. materials that they know better, such as Si and its compounds (Silicon Dioxide, Silicon Nitride) that can be integrated in the existing processes and that will not cause cross-contamination concerns in their ultra-clean production environments.

Whatever the approach for creating non-volatile memories, structural characterization is a critical part of the research effort that is required for their development. Because of the complicated nature of their functional properties (retention time, write/erase speed etc.), the structural and morphological data of these devices must be combined with other materials' properties and esp. the electrical, in order to understand their operation.

2 Flash Memories

Flash memories consist of at least 3 layers that are deposited on a semiconducting substrate (usually Si) covered by an electrode (or gate) where the write/erase voltage (VG) is applied (Fig. 1). In these 3 layers, the middle layer is the charge storage layer (floating gate or charge trap layer), where the charge is being stored and the other two layers are the insulating layers that block the transfer of electrons to and from the middle layer. The insulating layer next to the substrate (called "tunneling" layer) is thin enough to allow the control of the tunneling of the charges to the substrate, i.e. the charging and discharging of the charge storage layer; the tunneling layer is usually SiO_2 (thermally grown on the Si) because of the unique stability of this material, a crucial property for this position in the stack. The second insulating layer, the "blocking" or "control" oxide, has also very critical properties for the operation of the memory cell, as this is the medium through which the current may flow to the gate electrode. The device operates with the application of a voltage to the top electrode; charge is then injected from the substrate into the charge storage layer. The presence of charge near the Si substrate affects its local conductivity; if we replace the gate stack of a Field Effect Transistor (FET) with this stack, we have a transistor that can also store a bit permanently (see Fig. 1); it will

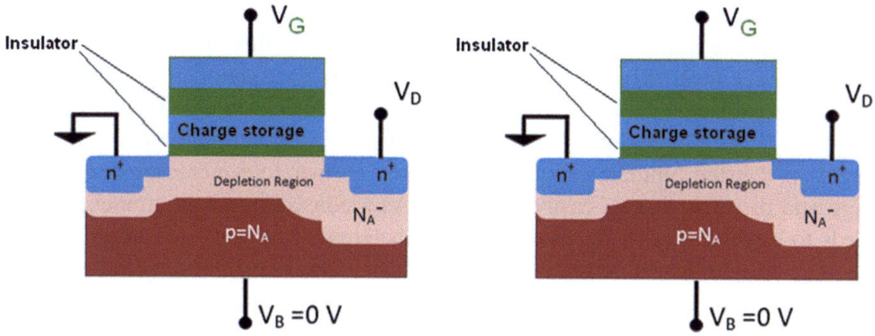

Fig. 1 The geometry of a Flash memory: *Blue* are the conducting areas, *green* are the insulators; on the right sketch there is charge stored near the channel so it is now conducting

not be erased until a voltage with an opposite sign is applied on the electrode. Until then, the charges will be stored there for a very long time (long "retention time").

If the charge storage layer is composed of a conducting material, then it is called "floating gate"; if it is an insulator, then it is called "charge-trap layer". The composition, the structure, the morphology, the crystallinity of the charge storage layer and the control layer have been an area of intensive research during the last few years, not only because we want to improve them for further miniaturization, but also because they can be engineered in order to tune the performance and the electrical characteristics of the memory cell, i.e. the amount of charge stored, the leakage currents, the charge retention time, the write/erase speed, its endurance etc.

We have worked on a large variety of materials and processes for the control and the charge storage layers. Because of the small volume of the layers, their structural characterization can be done by a limited number of techniques, with the TEM (Transmission Electron Microscopy) standing out of the rest. Of course, XRR (X-Ray Reflectivity), XPS (X-ray Photoelectron Spectroscopy), Grazing Incidence X-Ray Diffraction (GIXRD) provide also very important information; moreover they probe a much larger volume of the material than TEM.

The layers in our study are usually grown on full sheet wafers and are usually processed in relatively high temperatures (i.e. more than 200 °C), so the TEM samples can be thinned down to electron transparency by conventional means, that is, mechanical thinning followed by Ar ion milling at room temperature, without any effect on their structure. TEM of the layers was performed at plan view and cross sectional geometries with a Philips CM 20 HRTEM (High Resolution TEM) with a LaB_6 filament, equipped with an EDS (Energy Dispersive X-ray Spectroscopy) system and a GIF (Gatan Imaging Filter); the instrument is located at NCSR Demokritos. TEM has been able to provide the deeper inside view needed in this work and to resolve many open issues. Thickness, roughness, composition, intermixing of layers, defects (implantation damage etc.) and crystallite size, being the most critical, as these can affect seriously the electrical properties.

To overcome the problem of scaling of flash memories and to decrease the thickness of the tunneling oxide we worked on nanoparticle memories, where the charge storage layer (floating gate) is replaced by separate nanoparticles that are not in electric contact with each other. The reasoning behind this was that with conventional floating gates (single layers), the charging/discharging occurs through a single spot at the tunneling oxide; the oxide's temperature increases dramatically there and therefore it gets easily damaged. As a consequence, the tunneling oxide needs to be relatively thick. By fractioning the single floating gate into many parts we were able to distribute the conductive paths in the tunneling oxide over a wide area, and therefore reduce damages there. We managed to reduce the tunneling oxide thickness in a reliable way, and therefore to lower also the memory cell's power consumption. To achieve this we worked on the deposition of a large variety of metallic nanoparticles by e-gun evaporation on SiO_2/Si and covered them with a control oxide (SiO_2 or HfO_2). With the help of plan view TEM we managed to find the right parameters to create isolated self-assembled nanoparticles of various metals with a range of sizes. We used Pt [1–6], Pd [7], Au [4, 6, 8–10]. The TEM cross sections confirmed that the particles were located after processing on the interface between the control and the tunneling oxide and that they were covered completely by the control oxide; the retention time measurements and the other electrical characteristics confirmed the low voltage operation of these memory stacks and they showed that 1 or 2 electrons were stored per particle.

In another effort to create nanoparticle memories we used an innovative nanoparticle source to deposit Ni particles with much more accurate sizes than before, because the particles in this source are self-assembled in a low-vacuum pre-chamber. TEM played a crucial role in calibrating the operation of the nanoparticle source and confirmed the excellent size distribution of the particles, their purity, and their random distribution after their deposition, while the electrical measurements demonstrated their charge trapping properties. For Fig. 2 we deposited Ni particles on a Carbon support film used for TEM observation in order to avoid the Si thinning process and the related artifacts. For this particular type of particles, there is an oxide layer forming on the surface as it is clear on the diffraction contrast that arises; in actual memory stacks, Ni particles are embedded into the control layer immediately after their deposition, without any exposure to air [11, 12].

Besides charge storage in metals we used also a semiconductor, Si for this purpose. Amorphous Si was deposited by LPCVD (Low Pressure Chemical Vapor Deposition) on SiO_2/Si and was then annealed and oxidized. Si nanocrystals with various sizes and orientations were therefore embedded between 2 SiO_2 layers. As the crystals were very close to each other we were unable to confirm that they we not in contact, as the electrical measurements were suggesting. Nevertheless, this approach showed good charge trapping properties and a CMOS compatible method to create memory stacks [13, 14].

Charge trapping memory structures that use an insulator Si_3N_4 for their charge storage layer are now used in a number of applications, one example being the V-NAND structure; this is a 3D memory, currently on the market, introduced in 2013 by Samsung.

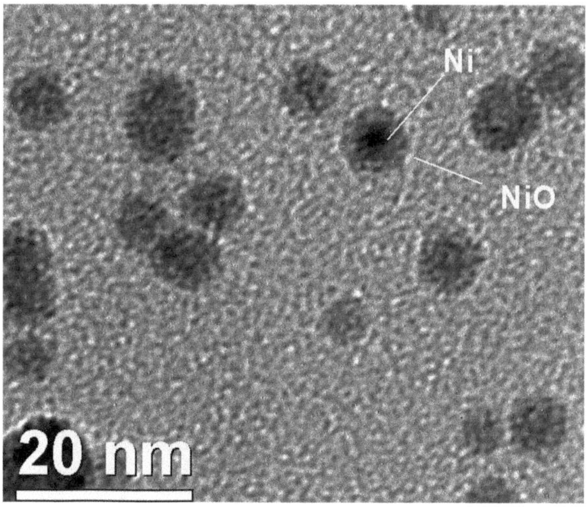

Fig. 2 Ni Nanoparticles deposited with a nanoparticle source

In order to engineer the properties of such memory cells SONOS (Silicon/silicon diOxide/silicon Nitride/silicon diOxide/Slicon) structures were fabricated by depositing Si_3N_4 on SiO_2/Si, followed by Si, N and Ar implantation and then by wet oxidation (to form the control oxide). The TEM and electrical characterization showed that the control oxide grows thicker on the Si-implanted samples and that the deep traps created by the implantation process provide an interesting option for such memories [15].

The replacement of the SiO_2 at the control oxide by alumina was studied with the growth of MANOS (Metal/Alumina/silicon Nitride/silicon Oxide/Silicon) structures. Alumina offers an alternative oxide to SiO_2, as it has larger band gap and permittivity.

Several processes for the optimization of these memories, see for example [16], were studied and especially those that affect the properties of the Alumina (Al_2O_3) layer. ALD (Atomic Layer Deposition) was used with trimethylaluminum (TMA) and water (or ozone) at 300 °C for the Al_2O_3 deposition; the structure was then annealed (PDA, Post Deposition Annealing) in nitrogen ambient. Without the PDA, Al_2O_3 is amorphous with very high leakage currents. PDA condensed the Al_2O_3 layer without any loss of mass, as confirmed by XRR and turned it into a poly-crystalline material (see Fig. 3); its crystallites increase with the PDA temperature. There is an interfacial layer between Al_2O_3 and Si_3N_4 which is formed right after the ALD deposition that is present also after the PDA. The leakage currents of the Al_2O_3 layer after PDA are reduced significantly.

Ion implantation of the ALD alumina control layer by low energy N atoms was another way of optimizing the function of the MANOS memory stacks. Amorphous alumina was implanted and then a post implantation annealing was performed (see Fig. 4). It was found that the implantation retards the crystallization of alumina. XPS data indicates that at the implanted area, Aluminum oxynitride is formed. By

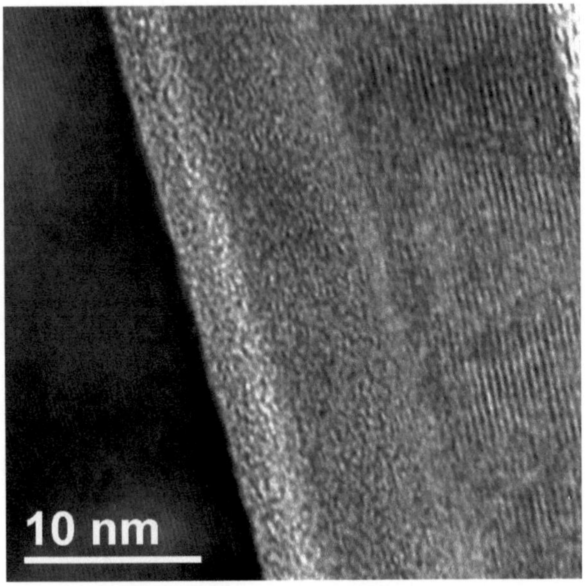

Fig. 3 MANOS stack after PDA

Fig. 4 HRTEM image of N ion beam implanted Al_2O_3/Si_3N_4/SiO_2/Si stack (at 1 keV, fluence 10^{16} ions/cm^2), Al_2O_3 was grown in ALD with TMA and water before the implantation. PDA: 1050 °C for 15 min

selecting the right implantation and annealing conditions the retention time of the memory stacks was improved by 25 %, when compared to the non-implanted layers.

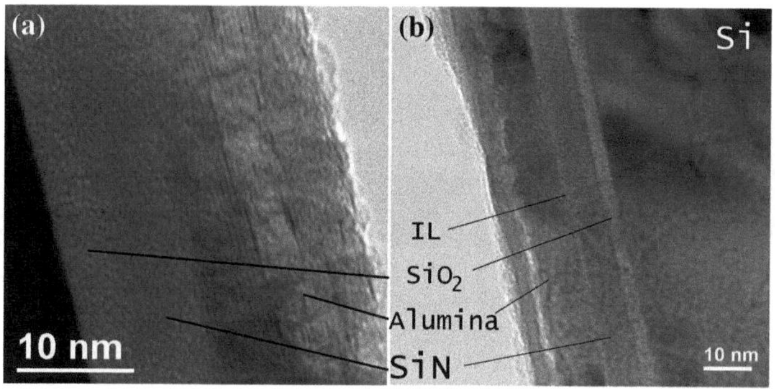

Fig. 5 **a** and **b** TEM images of N ion beam implanted $Al_2O_3/Si_3N_4/SiO_2/Si$ stack; Al_2O_3 has been grown in ALD with TMA and ozone. PDA: 850 °C for 15 min

In Fig. 5 we can notice the interfacial layer (IL) between the alumina and the nitride layer as well as the large variation in the position of the implanted volume (the area with the smaller density inside the alumina); we can see also the presence of polycrystalline areas around it.

Besides Al_2O_3, other high-k materials were deposited by ALD for the replacement of the SiO_2 of the control oxide over the $Si_3N_4/SiO_2/Si$; these were ZrO_2 and HfO_2 [17–19]. An improved write/erase performance (>3 × 10^5, in the 10/−11 V, 10 ms regime) was achieved by the appropriate selection of the ALD precursors; this meant that different deposition temperatures were used that affected the nature of the Si_3N_4/HfO_2 and the Si_3N_4/ZrO_2 interfaces.

3 Resistive Memories

Resistive memories have a much simpler structures than charge-trap memories. They usually consist of 2 electrodes that are separated by a sub-stoichiometric material, such as an oxide, e.g., TiO_{2-x}. When applying sufficiently high voltages the material enters a low resistivity state; when applying a voltage of the opposite polarity the material returns to a high resistivity state; such materials can switch between resistivity states thousands of times. The exact nature of this phenomenon is not yet clearly understood with a large part of the scientific community speculating that it is a result of filamentary vacancy migration.

This does not stops semiconductor companies from introducing the first Resistive RAMs (Random Access Memories) to the market, with very low power consumption. This technology has a huge potential for commercialization, also because its simplicity means relatively simple growth processes and patterning.

CAFM (Conducting Atomic Force Microscopy) is a technique that can simultaneously measure the surface structure of a layer and its relative resistivity. In these

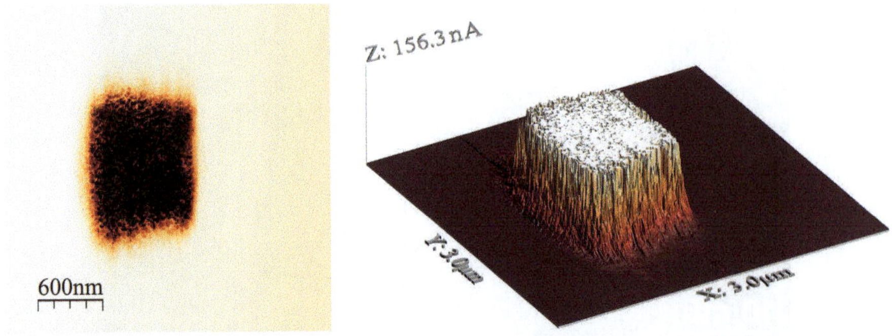

Fig. 6 C-AFM image (*left*) and its 3D representation (*right*) taken with a tip voltage of +4 V (READ). The rectangular area is the low resistivity area; it was scanned (before) with 7 V (WRITE)

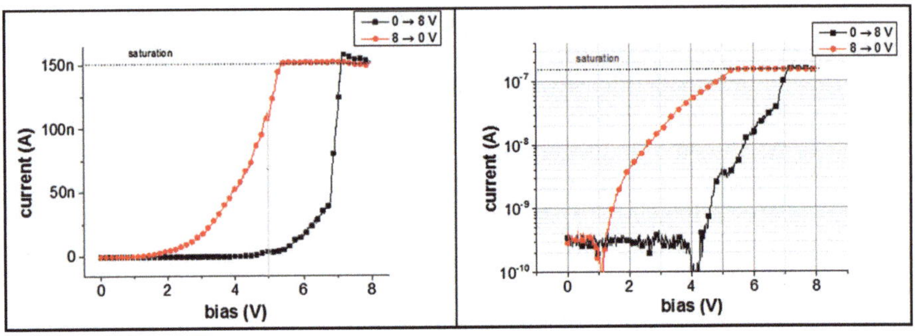

Fig. 7 I-V curve taken on the CAFM; Voltage is on the tip, the sample is grounded

materials it has been particularly helpful in the study of Ti (4 nm)/TiO_{2-x} (45 nm) layers that were sputtered on Au/SiO_2/Si.

CAFM was performed with a Veeco CP II AFM equipped with a DLPCA-200 low-noise linear amplifier by Femto. Bruker's SCM PIT tip (Pt/Ir coated, nominal radius of 35 nm, spring constant: 3 N/m), gave the best results among the 3 tips that we used.

In Fig. 6 we see a demonstration of a way to write and read a resistive memory with the CAFM. Note that the topographic images are not influenced by the write/erase process. Figure 7 shows an I-V curve recorded on certain spots, where the CAFM tip was resting; they demonstrate an on/off ratio of about 100; in our attempts to image with CAFM any of these isolated spots we were not able to locate them. When we "wrote" on several nearby spots we managed to image the low resistive area on the CAFM. This indicates that the low resistive area that is formed by the tip is smaller than the resolution of the CAFM measurements. Therefore we have a direct indication that in this particular material our tip writes an area smaller than ~50 nm. This can be critical for the scaling of memory cells.

References

1. C. Sargentis, K. Giannakopoulos, A. Travlos, N. Boukos, D. Tsamakis, Appl. Phys. Lett. **88**(7), 073106–073106-3 (2006)
2. C. Sargentis, K. Giannakopoulos, A. Travlos, D. Tsamakis, Surf. Sci. **601**(13), 2859–2863 (2007)
3. C. Sargentis, K. Giannakopoulos, A. Travlos, D. Tsamakis, in *MRS Proceedings*, vol. 830, D6 4 (2004)
4. C. Sargentis, K. Giannakopoulos, A. Travlos, D. Tsamakis, in *International Semiconductor Device Research Symposium*, pp. 1–2, (2007)
5. C. Sargentis, K. Giannakopoulos, A. Travlos, D. Tsamakis, J. Phys. Conf. Ser. **10**(1), 53 (2005)
6. C. Sargentis, K. Giannakopoulos, A. Travlos, D. Tsamakis, Mater. Sci. Semicond. Process. **12**(1), 57–63 (2009)
7. C. Sargentis, K. Giannakopoulos, A. Travlos, D. Tsamakis, in *International Semiconductor Device Research Symposium*, pp. 218–219 (2003)
8. C. Sargentis, K. Giannakopoulos, A. Travlos, D. Tsamakis, Physica E **38**(1), 85–88 (2007)
9. C. Sargentis, K. Giannakopoulos, A. Travlos, P. Normand, D. Tsamakis, Superlattices Microstruct. **44**(4), 483–488 (2008)
10. C.H. Sargentis, K. Giannakopoulos, A. Travlos, D. Tsamakis, G. Krokidis, in *International Semiconductor Device Research Symposium*, pp. 342–343 (2005)
11. E. Verrelli, D. Tsoukalas, K. Giannakopoulos, D. Kouvatsos, P. Normand, D.E. Ioannou, Microelectron. Eng. **84**(9), 1994–1997 (2007)
12. E. Verrelli, D. Tsoukalas, K. Giannakopoulos, D. Ioannou, in *MRS Proceedings*, vol. 997, 0997-I03-08 (2007)
13. E. Tsoi, P. Normand, A.G. Nassiopoulou, V. Ioannou-Sougleridis, A. Salonidou, K. Giannakopoulos, J. Phys: Conf. Ser. **10**(1), 31 (2005)
14. A. Salonidou, A.G. Nassiopoulou, K. Giannakopoulos, A. Travlos, V. Ioannou-Sougleridis, E. Tsoi, Nanotechnol. **15**(9), 1233 (2004)
15. D. Simatos, P. Dimitrakis, V. Ioannou-Sougleridis, P. Normand, K. Giannakopoulos, B. Pecassou, G. BenAssayag, in *MRS Proceedings*, vol. 1430, mrss12-1430-e03-01 (2012)
16. N. Nikolaou, V. Ioannou-Sougleridis, P. Dimitrakis, P. Normand, D. Skarlatos, K. Giannakopoulos, K. Kukli, J. Niinistö, M. Ritala, M. Leskelä, Thin Solid Films **533**, 5–8 (2013)
17. N. Nikolaou, P. Dimitrakis, P. Normand, V. Ioannou-Sougleridis, K. Giannakopoulos, K. Mergia, K. Kukli, J. Niinistö, M. Ritala, M. Leskelä, Solid-State Electron **68**, 38–47 (2012)
18. N. Nikolaou, P. Dimitrakis, P. Normand, V. Ioannou-Sougleridis, K. Giannakopoulos, K. Mergia, K. Kukli, J. Niinisto, M. Ritala, M. Leskela, Adv. Mater. Res. **324**, 42–45 (2011)
19. V. Ioannou-Sougleridis, N. Nikolaou, P. Dimitrakis, P. Normand, K. Giannakopoulos, K. Mergia, K. Kukli, J. Niinisto, M. Ritala, M. Leskela, in *Materials Research Society Symposium Proceedings* vol. 1250 (2010)

Advanced Technology for Analytical Electron Microscopy by Using Aberration Corrected Transmission Electron

Guillaume Brunetti

Abstract This manuscript is devoted to the presentation of the last High-End Transmission Electron Microscope (TEM) developed by JEOL: the JEM-ARM200F equipped with a Cold Field Emission Gun (CFEG) and Cs S/TEM correctors. Application examples will be presented.

The nanoworld: where individual atoms that constitute substances are directly observed. A TEM equipped with Cold-FEG (Cold Field Emission Gun) realizes this dream. With Cold-FEG, a superb high-quality electron beam is produced that achieved a narrower spread and forms a sharper probe with higher brightness than a conventional Schottky FEG. Many improvements have been developed for the conception of the ARM200F.

The ARM200F (Fig. 1) receives a new design allowing the improvement of the mechanical, electrical stability and the protection against environmental disturbances. Furthermore, the base frame has been designed to the optimum geometry for the accommodation of double TEM/STEM Cs correctors totally integrated. The use of a spherical aberration corrector for electron optic system as standard, has achieved a scanning transmission image (STEM-HAADF) resolution of 0.078 nm, the highest in the world among the commercial transmission electron microscopes.

These improvements coupled with Cold-FEG allow the ARM to perform high level studies for all conventional TEM techniques (HRTEM, HRTEM, EELS, EDS, Annular Bright Field, diffraction …) and also more specific techniques (Holography, precession, …). Furthermore, the ARM200F equipped with Cold-FEG and the new Centurio EDS detector (solid angle 1 sr) allows the acquisition of atomic resolution EDS mapping (Fig. 2).

G. Brunetti (✉)
JEOL (Europe) SAS—Espace Claude Monet, 1 Allée de Giverny,
78290 Croissy-sur-Seine, France
e-mail: brunetti@jeol.fr

E.K. Polychroniadis et al. (eds.), *2nd International Multidisciplinary Microscopy and Microanalysis Congress*, Springer Proceedings in Physics 164,
DOI 10.1007/978-3-319-16919-4_3

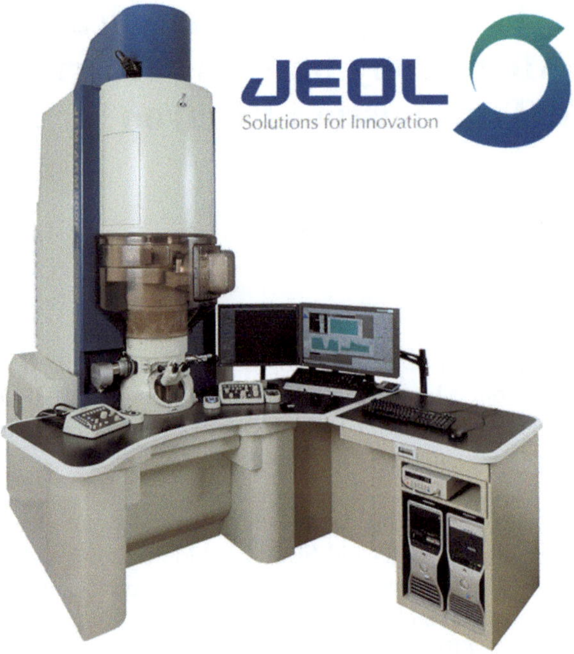

Fig. 1 JEM-ARM200F equipped with Cold Field Emission Gun

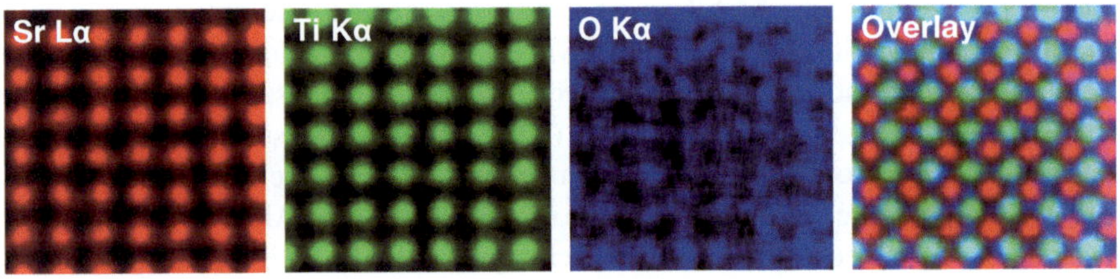

Fig. 2 Atomic resolution EDS mapping on $SrTiO_3$ sample

Examples of application:

(1) The use of a Cold-FEG dramatically improves EELS energy resolution. As seen Fig. 3, the energy resolution is about 0.3 eV for a Cold-FEG. This very good energy resolution is a major factor for EELS studies; in particular ELNES (Energy Loss Near Edge Structure) exhibits a characteristic shape depending on the chemical bonding states in a substance. Figure 3 is an example of rutile and anatase analysis.
(2) Annular Bright Field (ABF) technique was developed by JEOL for imaging light atoms column like Hydrogen, Oxygen, Lithium, … This technique is available as a standard method with ARM-200F. ABF is a powerful technique

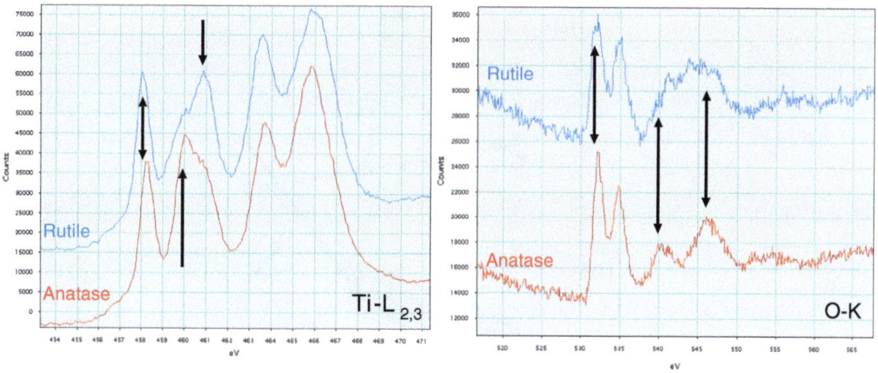

Fig. 3 EELS spectra of rutile and anatase type crystal, which show ELNES obtained from the O–K edge ans the Ti–L edge, reveal clear differences in chemical bonding states (indicated by *arrows*). The higher energy resolution of a Cold-FEG enables clear observation of these differences

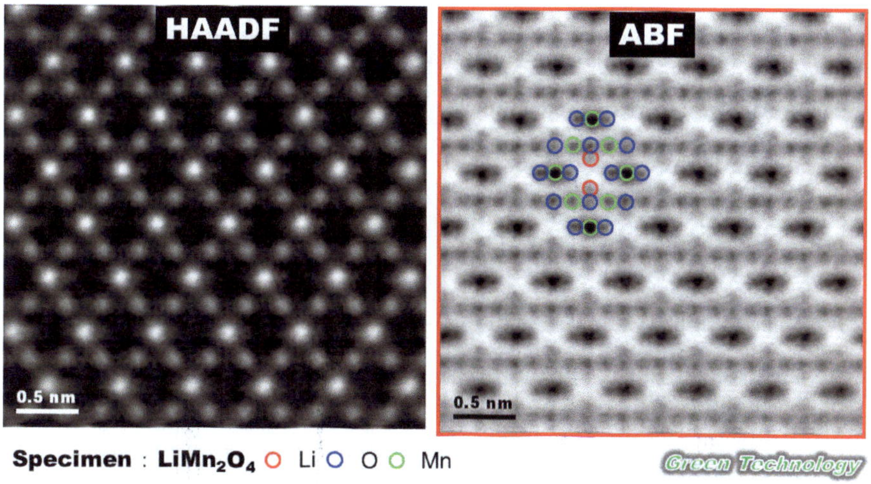

Fig. 4 Imaging of lithium by ABF. The lithium-atom columns are not visible with HAADF technique (**a**). With ABF, the lithium-atom columns are clearly visible (**b**). Sample $LiMn_2O_4$

for studying lithium battery (Fig. 4). Another application has been performed by Ishikawa et al. with the study of Hydrogen-atom columns [1].

(3) Zinc oxide (ZnO) materials are used in a number of technical applications and modification with transition metals can lead to novel combinations of properties e.g. semiconductivity and ferromagnetism. Doping with trivalent metal ions (e.g. In3+ or Fe3+) results in characteristic inversion domain structures. Domains in ZnO doped with In_2O_3 or Fe_2O_3 are separated by two types of inversion domain boundaries (IDBs) (Fig. 5).

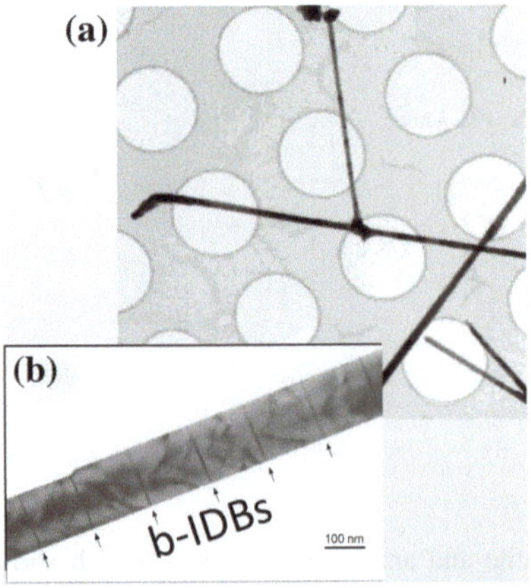

Fig. 5 Structure of In–ZnO nanorods. **a** Nanorods on carbon film (TEM-BF). **b** Section of In–ZnO nanorod showing characteristic basal b-IDBs (marked by *arrows*) viewed in edge-on orientation (Z.A. [$1\bar{1}20$]); pyramidal p-IDBs are inclined and thus are not clearly imaged

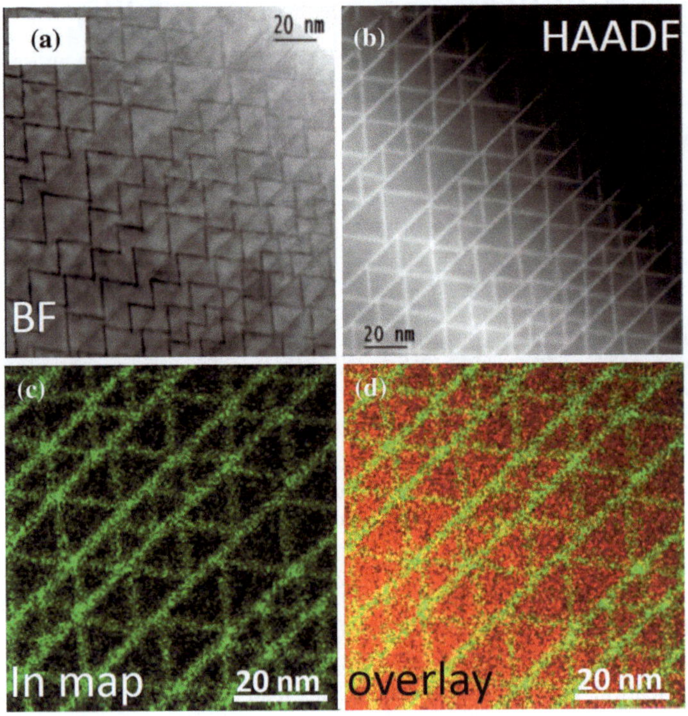

Fig. 6 STEM structural and spectroscopic imaging by X-ray (SIX). Basal b-IDBs and pyramidal p-IDBs viewed edge-on in [$1\bar{1}00$] projection (**a–d**). Both IDBs containing monolayers of In are visualized by BF (**a**), HAADF (**b**) STEM imaging, as well as EDS maping (**c**); corresponding overlay shown in (**d**), (In *green*; Zn *red*)

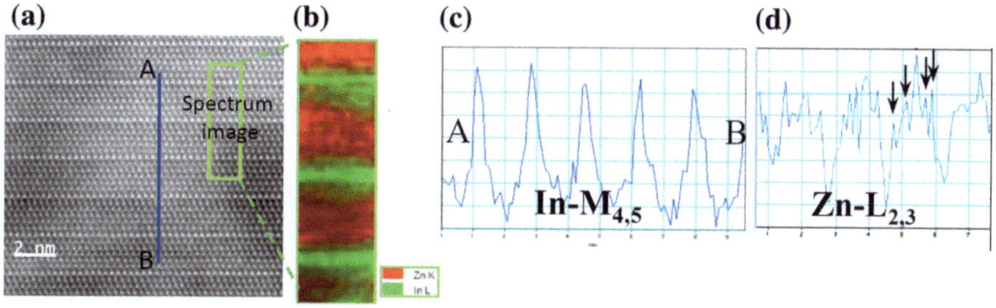

Fig. 7 HAADF STEM image of In–ZnO nanorod (**a**). Regions of interest for ESI are indicated: **b** EEL spectrum image (In *green*; Zn *red*); line-scan signals for In–M (**c**) and Zn–L (**d**). Remarkably, single Zn (0002) planes were resolved (*arrows* in **d**)

Evolving defect structures were characterized by high-resolution structural and spectroscopic imaging in the Cs-corrected JEOL JEM-ARM200F TEM/STEM (Figs. 6 and 7) [2].

With ARM-200F, JEOL offers "Ultra-high resolution" with this next generation of "improved Cold-FEG".

References

1. R. Ishikawa, E. Okunishi, H. Sawada, Y. Kondo, F. Hosokawa, E. Abe, Direct imaging of hydrogen-atom columns in a crystal by annular bright-field electron microscopy. Nat. Mater. **10**, 278–281 (2011)
2. JEOL Application Letter 05-2014, Characterization of inversion domains in ZnO nano materials with JEOL JEM-ARM200F

TEM Characterization of a Complex Twinning System in 3C–SiC

Mamour Sall, Narendraraj Chandran, Anastasia Terzidou, Christos B. Lioutas and E.K. Polychroniadis

Abstract This work was performed on a SiC sample grown on a 4H–SiC seed by chemical vapour deposition with the addition of GeH_4 gas to the classical SiH_4+C_3H_8 precursor system. In the present work, the defect structure and the polytypes "competition" are described. A complex 3C twinning system is also studied. This V-shape defect is nucleated inside the 3C layer contrary to what is generally observed (nucleation of twin at the 3C/4H interface). Using TEM diffraction, we identified the V-shape defect as a fourfold twin configuring three twin planes. Two of them are {111} twins, though the third one has been identified as a (11-5) twin.

1 Introduction

For the last decades, substantial research has been focused on the development of Silicon Carbide (SiC) electronic devices thanks to its outperformance over Si. For example, 4H–SiC based devices have been mainly developed due to its relative superior properties over other SiC polytypes. On the contrary, 3C SiC, despite the fact it is more advantageous for MOSFETs with higher channel mobility [1–3], is

M. Sall (✉) · N. Chandran · A. Terzidou · C.B. Lioutas · E.K. Polychroniadis
Department of Physics, Aristotle University of Thessaloniki, 54124 Thessaloniki, Greece
e-mail: mamour.sall@gmail.com

N. Chandran
e-mail: naren@physics.auth.gr

A. Terzidou
e-mail: anterzid@physics.auth.gr

C.B. Lioutas
e-mail: lioutas@physics.auth.gr

E.K. Polychroniadis
e-mail: polychr@auth.gr

E.K. Polychroniadis et al. (eds.), *2nd International Multidisciplinary Microscopy and Microanalysis Congress*, Springer Proceedings in Physics 164,
DOI 10.1007/978-3-319-16919-4_4

not used in today's electrical applications. This is due to the high density of defects, especially when Si is used as substrate because of the large lattice and thermal expansion coefficient mismatches between the materials [4]. Better matching can be obtained when using α-SiC (4H or 6H) as a substrate for 3C–SiC growth [5–7]. But experimentally, the growth of high quality 3C–SiC (111) layers on α-SiC (0001) substrates is very difficult due to the quasi-systematic formation of a high density of twin boundaries (TB). Recently, a chemical vapour deposition (CVD) technique using a GeH_4 gas flow as a surface pre-treatment prior to the growth, has demonstrated the possibility to drastically reduce twin boundaries in 3C–SiC grown on 4H SiC substrate with an optimal GeH_4 flux [8]. At particular GeH_4 fluxes, a competition between homo and heteroepitaxy has been observed. Additionally, the growth of 4H and 3C–SiC may promote the creation of twin boundaries in the 3C layer. In the present work, the formation of a complex twinning system inside the 3C–SiC is studied.

2 Experiment

The studied sample has been grown by chemical vapour deposition (CVD) with the addition of GeH_4 gas to the classical SiH_4+C_3H_8 precursor system. For TEM observations, cross-section sample ([11–20] orientation) is prepared by conventional mechanical thinning down to 50 μm. The electron transparency is achieved by ion milling using 4 keV Ar (Gatan PIPS). The selected-area electron diffraction (SAED) and TEM imaging were carried out in a 200 kV JEOL 2000 microscope.

3 Results and Discussion

Figure 1a, b shows representative cross section TEM images of the sample. A high competition between 3C–SiC heteroepitaxy and 4H–SiC homoepitaxy can be observed. That is illustrated in the schematic representation in Fig. 2. A clear mixing of 4H and 3C occurs near the 4H substrate. But after a certain thickness (near 1 μm), only 3C subsists.

Besides this growth competition (3C and 4H), we observe the formation of a complex twinning system with a formation of a V-shape defect implying some roughness of the surface of the sample (Fig. 1b). This defect is generated within the 3C layer unlike to what is generally observed (nucleation at the interface 3C/4H [9]. By recording the SAED around and inside the V-shape defect, we have identified

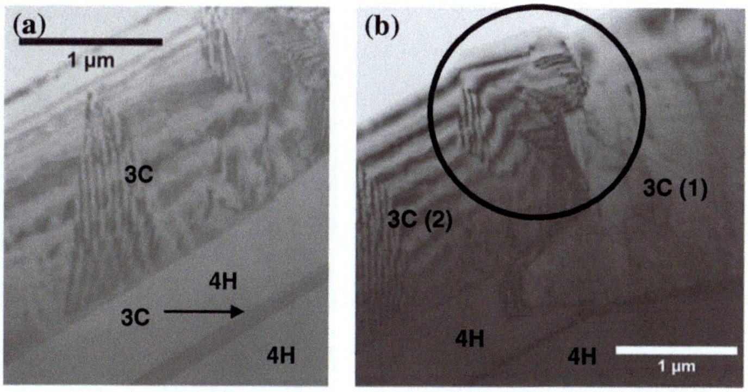

Fig. 1 Representative cross-section conventional TEM images showing the high competition between 3C and 4H growth. The *circle* in Fig. 1b indicates the V-shaped complex twinning system discussed in this section

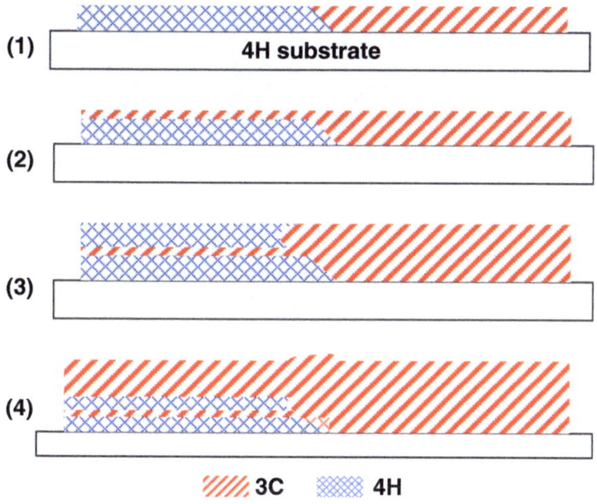

Fig. 2 Schematic illustration of the competition between 3C heteroepitaxy and 4H homoepitaxy. From (1) to (2), 3C and 4H layers grown from each part of the crystal meet, and then the 3C growth dominates. But in (3), the 3C growth is partially interrupted by 4H growth. And finally in (4), the 4H growth is stopped, and only 3C grows

four different 3C orientations (3C (1), 3C (2), 3C (3) and 3C (4) in Fig. 3a, b). As it can be seen in the diffraction pattern in Fig. 3b, these crystals make a fourfold twin configuring three twin planes. Two of them have the common {111} twin plane, though the third one has been identified as a (11-5) twin.

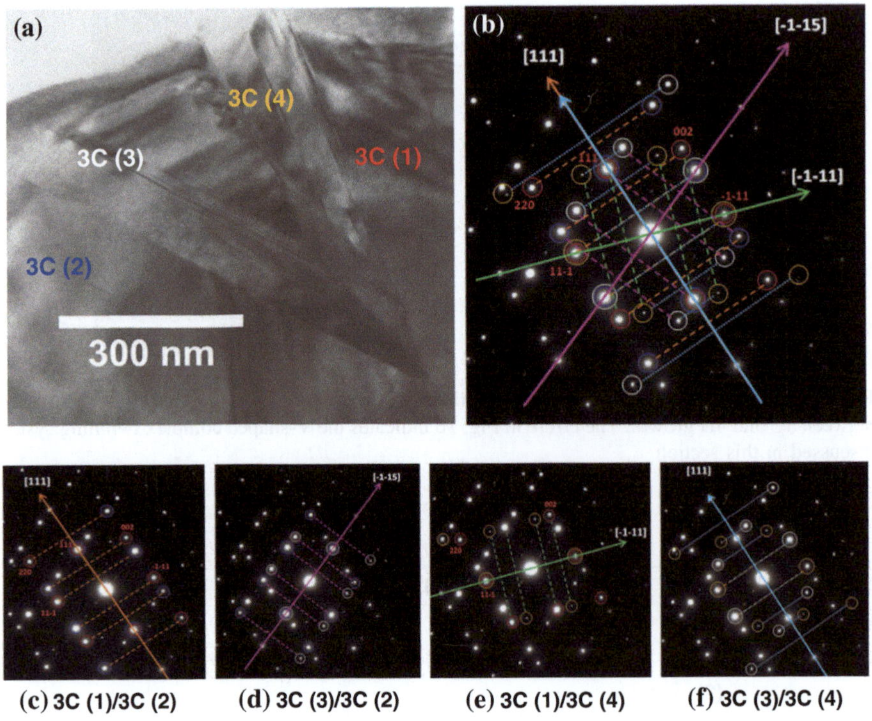

Fig. 3 **a** On-axis TEM image of the V-shape defect. The study of the diffraction pattern (**b**) recorded from this region has shown the presence of four 3C orientations establishing three twinning relations along the common {111} planes and a particular one along (11-5) plane. The detailed study of the same DP is represented in (**c–f**) lighting up the relation between the different 3C orientations. All the indices are defined based on a unique crystallographic orientation with 3C (1) as reference

4 Conclusion

We have studied by means of TEM a 3C–SiC sample grown on a 4H–SiC seed by CVD with the addition of GeH_4 gas to the classical SiH_4+C_3H_8 precursor system. The particular used GeH_4 flux for this sample has led to a high competition between 3C–SiC and 4H–SiC growths in the beginning of the growth. The 3C layer alone appears only after 1 µm growth. Additionally within the 3C layer, we have shown a fourfold twin V-shape defect configuring three twin planes: the common {111} twin planes and a (11-5) twin plane.

Acknowledgments This work is financially supported by Marie Curie Actions under the project no 264613-NetFISiC. The authors are indebted to Dr G. Ferro and K. Alassad from UCBL 1 Lyon for having provided the studied sample.

References

1. W.E. Nelson, F.A. Halden, A. Rosengreen, Growth and properties of β-SiC single crystals. J. Appl. Phys. **37**, 333–336 (1966)
2. H. Nagasawa, M. Abe, K. Yagi, T. Kawahara, N. Hatta, Fabrication of high performance 3C-SiC vertical MOSFETs by reducing planar defects. Physica Status Solidi B **245**, 1272–1280 (2008)
3. Y.H. Zhu, J.C. Zhang, Z.T. Chen, T. Egawa, Demonstration on Gan-based light-emitting diodes grown on 3C-SiC/Si(111). J. Appl. Phys. **106**, 124506 (2009)
4. A. Severino, C. Locke, R. Anzalone, M. Camarda, N. Piluso, A.L. Magna, S. Saddow, G. Abbondanza, G. D'Arrigo, F.L. Via, Growth of wide bandgap materials. ECS Trans. **35**, 99 (2011)
5. K. Nishino, T. Kimoto, H. Matsunami, Reduction of double positioning twinning in 3C-SiC grown on α-SiC substrates. Jap. J. Appl. Phys. **36**, 5202
6. M.V.S. Chandrashekhar, C.I. Thomas, J. Lu, M.G. Spencer, Electronic properties of a 3C/4H SiC polytype heterojunction formed on the Si face. Appl. Phys. Lett. **90**, 173509 (2007)
7. M. Soueidan, G. Ferro, O. Kim-Hak, F. Cauwet, B. Nsouli, Vapor–liquid–solid growth of 3C-SiC on α-SiC Substrates. 1. Growth mechanism. Cryst. Growth Des. **8**, 1044–1050 (2008)
8. K. Alassasd, M. Vivona, V. Soulière, B. Doisneau, F. Cauwet, D. Chaussende, F. Giannazzo, F. Roccaforte, G. Ferro, Ge mediated surface preparation for twin free 3C-SiC nucleation and growth on low off-axis 4H-SiC substrate. ECS J. Solid State Sci. Technol. **8**, 285 (2014)
9. M. Marinova, A. Mantzari, E.K. Polychroniadis, Some recent results on the 3C-SiC structural defects. Solid State Phenom. **159**, 39 (2010)

Microstructural Evaluation of Suspension Thermally Sprayed WC-Co Nanocomposite Coatings

R. Ahmed, N.H. Faisal, O. Ali, Nayef M. Al-Anazi, S. Al-Mutairi, S. Mamour, E.K. Polychroniadis and M.F.A. Goosen

Abstract Microstructural and sliding wear evaluations of nanostructured coatings deposited by Suspension High Velocity Oxy-Fuel (S-HVOF) spraying were conducted in as-sprayed and HIPed (Hot Isostatically Pressed) conditions. S-HVOF coatings were nanostructured via ball milling of the WC-12Co start powder, and deposited via an aqueous based suspension using modified HVOF (TopGun) spraying. Microstructural evaluations of these hardmetal coatings included TEM (Transmission Electron Microscopy), X-ray Diffraction (XRD) and Scanning Electron Microscopy (SEM). Sliding wear tests were conducted using a ball-on-flat test rig. Results indicated that nanostructured features inherited from the start powder in S-HVOF spraying were retained in the resulting coatings. The decarburisation of WC due to a higher surface area to volume ratio was also observed in the S-HVOF coatings. Nanostructured and amorphous phases caused by the high

R. Ahmed (✉) · N.H. Faisal
College of Engineering, Alfaisal University, P.O. Box 50927, Riyadh 11533, Saudi Arabia
e-mail: R.Ahmed@hw.ac.uk

R. Ahmed · O. Ali
School of Engineering and Physical Sciences, Heriot-Watt University, Edinburgh EH14 4AS, UK

N.H. Faisal
School of Engineering, Robert Gordon University, Garthdee Road, Aberdeen AB10 7GJ, UK

N.M. Al-Anazi · S. Al-Mutairi
Materials Performance Unit, Research & Development Centre, Saudi Aramco, Dhahran 31311, Saudi Arabia

S. Mamour · E.K. Polychroniadis
Department of Physics, Aristotle University of Thessaloniki, 54124 Thessaloniki, Greece

M.F.A. Goosen
Office of Research & Graduate Studies, Alfaisal University, P.O. Box 50927, Riyadh 11533, Saudi Arabia

E.K. Polychroniadis et al. (eds.), *2nd International Multidisciplinary Microscopy and Microanalysis Congress*, Springer Proceedings in Physics 164,
DOI 10.1007/978-3-319-16919-4_5

cooling rates during thermal spraying crystallized into complex eta-phases after the HIPing treatment. Sliding wear performance indicated that the coating wear was lower for the HIPed coatings.

1 Introduction

Thermally sprayed WC-Co coatings are used for a variety of tribological applications in aerospace, transportation, energy and off-shore sectors [1, 2]. The WC contributes to the high hardness whereas Co provides toughness to the deposited coating. Conventionally, these coatings are sprayed using agglomerated and sintered WC-Co powder particles in the typical size range of −45/+5 μm [1–3]. A recent advancement is to consider nano-composite powder particles which are deposited using an aqueous or ethanol suspension [1–4]. The aim of this paper was to consider further possible improvement of these suspension nano-composite coatings via post-treatment using Hot Isostatic Pressing (HIPing). Results are discussed in terms of the microstructural and tribological changes occurring as a result of HIPing.

2 Experimental

2.1 Coating Deposition and Post-treatment

Suspension High Velocity Oxy-Fuel (S-HVOF) spraying was conducted using a modified HVOF (TopGun, GTV mbH, Luckenbach, Germany) spray process using ethene as the fuel gas [4]. The construction of the modified torch allows an axial injection of the suspension into the burning chamber. An agglomerated and sintered WC-12 wt% Co spray feedstock powder (Fujimi Corp., Japan—DTS W653-20/5) with sub-micron WC grains was selected for the spray process. In order to adapt the particle size for suspension preparation, the powder was milled in a planetary ball mill and the resulting nanocomposite powder was used to produce aqueous suspension for S-HVOF spraying. Table 1 summarizes the coating process parameters used for coating deposition. Further details of the S-HVOF spraying process can be seen in earlier publications [4]. After spraying the coatings were heat treated (Hot Isostatically Pressed (HIPed) at 920 °C for 2 h at 103 MPa in an inert atmosphere) for the transformation of the amorphous constituents into a crystalline state.

Coatings were also analyzed for their total carbon content by the combustion method (CS 230, LECO Corporation, St. Joseph, MI, USA) and for oxygen content by the carrier gas hot extraction technique (TCH 600, LECO Corporation, St.

Table 1 S-HVOF spray parameters [4]

C_2H_4 (l/min)	O_2 (l/min)	Spray distance (mm)
75	170	90

Joseph, MI, USA). All coatings were deposited on AISI440C steel discs of 31 mm diameter and 6 mm thickness. The surface of as-deposited coatings was ground and polished prior to tribological wear testing.

2.2 *Microstructural Evaluation*

The microstructure of the coatings was observed via Scanning Electron Microscopy (SEM) equipped with Secondary Electron (SEI) and Backscattered Electron Imaging (BEI). The chemical compositions of the microstructural phases in the powders and the coatings were determined via X-ray Diffraction (XRD) with Cu–Kα radiation (wavelength = 1.5406 Å). Transmission Electron Microscopy (TEM) observations of the samples were made on the surface of the coating layers. The samples were prepared under plane view configuration. They were thinned down to less than 50 μm by mechanical polishing. The electron transparency was achieved by ion milling at 5 kV using the GATAN precision ion polisher system (PIPS) at an incidence angle of 5°. The observations were performed on a conventional JEOL 2000 TEM operating at 200 kV.

2.3 *Sliding Wear Investigations*

The sliding wear resistance was examined via ball-on-flat tests at room temperature on a bench mounted wear test machine (BLR2000M; Bud Labs, USA). The ball-on-flat tests were conducted using AISI 440C steel ball and coated disc sample under a normal load of 25 N. The test conditions were similar to ASTM G133-02 (procedure A), except that the ball radius was slightly larger at 6.35 mm and sliding distance 500 m. During the test, the disc experienced reciprocating sliding motion at an oscillating frequency of 2.0 Hz, with a stroke length of 10 mm. Five tests were conducted for each test couple. Wear volume loss of the coating was computed from the length of the stroke and the average cross-sectional area of the wear grooves, which was measured via the interferometer (Zygo New View). The friction coefficient was evaluated using a tension-compression load cell mounted on the sliding wear rig. Averaged friction coefficient values and their standard deviations are presented in the results section.

3 Results

Figure 1 shows the SEM comparison of the milled powder, as-sprayed and HIPed S-HVOF coatings. The XRD comparison of the start powder, as-sprayed and HIPed coatings is shown in Fig. 2. Figure 3 and Table 2 shows the TEM investigation

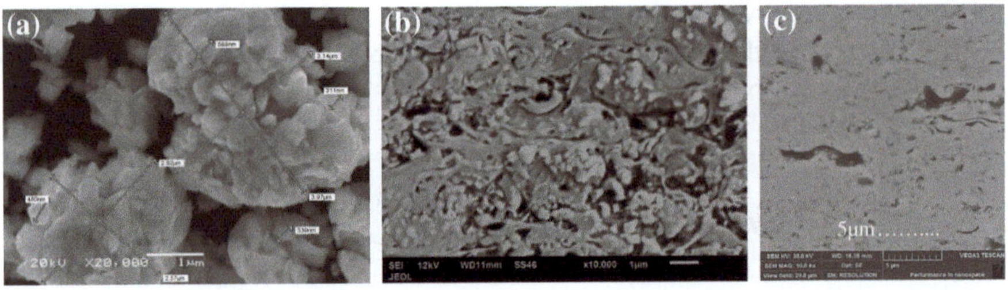

Fig. 1 SEM observations of **a** milled powder, **b** as-sprayed and **c** HIPed S-HVOF coatings

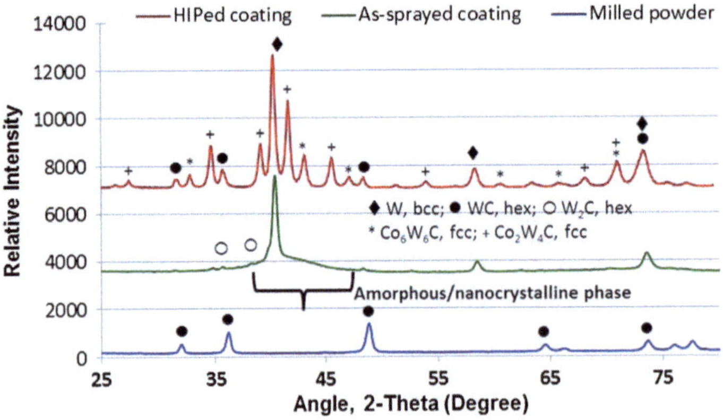

Fig. 2 XRD comparison of milled powder, as-sprayed and HIPed S-HVOF coatings

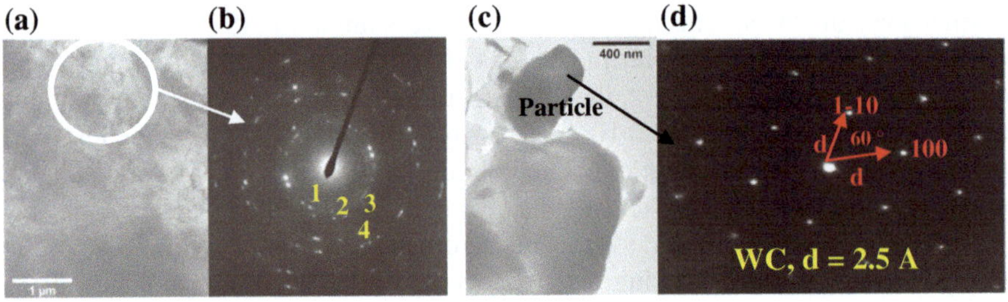

Fig. 3 TEM observations of as-sprayed S-HVOF coating. **a**, **b** Polycrystalline region, the different studied diffraction rings are indicated. **c**, **d** Single crystal particle identified as WC

analysis of the as-sprayed coating. The TEM observations of the HIPed coating are shown in Fig. 4. The analysis of diffraction rings for the HIPed coating is presented in Table 3. The non-metal analysis for the as-sprayed coating indicated that the carbon and oxygen content was 1.2 and 1.0 %, respectively. Figure 5a summarizes the total (ball and coating) wear rate comparison of the as-sprayed and HIPed coating. The friction plot comparison of the two coatings is shown in Fig. 5b.

Table 2 Experimental d-spacing measured for different rings and theoretical d-spacing for some possible phases for the as-sprayed coating

		d_{th} (A)						
Ring number	d_{exp} (A)	WC (hex.)	W_2C (hex.)	W (cubic)	Co_3W_3C (cubic)	Co_6W_6C (cubic)	$Co_3W_9C_4$ (hex.)	Co (hex.)
1	2.2 (weak)	–	2.3 (−1−11) Int100	2.2 (011) Int100	2.1 (511) Int100	2.1 (511) Int100	2.2 (301) Int100	2.2 (010) Int100
2	1.9 (strong)	1.9 (101) Int83	–	–	1.9 (440) Int40	1.9 (440) Int35	1.9 (302) Int10	1.9 (002) Int25
3	1.7	–	1.7 (−1−12) Int15	1.6 (002) Int17	1.7 (620) Int2	–	1.7 (400) Int11	–
4	1.2	1.2 (201) Int11	1.1 (−2−22) Int3	1.1 (022) Int12	1.1 (755) Int13	–	1.1 (325) Int8	1.2 (013) Int16

In brackets are the hkl values. The intensities of the X-ray reflections are also indicated in the table

4 Discussion

4.1 Coating Microstructure

The surface morphology of the coating as revealed in Fig. 1 indicated lamella microstructure with low porosity coatings which is typical of thermal spray coatings [1–3]. The nanostructured features observed in Figs. 1b and 2 originated from the milled powder (Fig. 1a). The TEM analysis of as-sprayed coating (Fig. 3 and Table 2) indicated the presence of nanosized WC, W, Co_3W_3C and crystalline Co. For discussion purposes, these particles in the deposited coating will be collectively termed as "nanosized particles". TEM analysis indicated that these nanosized particles were well distributed within the coating microstructure.

The XRD patterns of the as-sprayed coatings indicated WC, W and some W_2C with shifted peak positions, but the pattern was dominated by a large "nanocrystalline peak" between 2θ values of 35–48°. The XRD patterns of the milled spray powder showed only tungsten carbide (WC) peaks (Fig. 2). The S-HVOF deposition process led to intensive structural changes in the material which is consistent with the previous studies by Berghaus et al. [2]. No metallic Co was observed in the XRD pattern of coatings suggesting that it became part of the amorphous/nanocrystalline matrix. Therefore, an amorphous or nanocrystalline binder phase was produced as observed in Fig. 2. The decarburisation of WC was due to a higher surface area to volume ratio of the powder particles in the S-HVOF coatings.

The carbon analysis of the coating indicated an intensive carbon loss. At the same time the oxygen content was also higher than usually observed for WC-Co coatings (typically <0.2 wt% [4]). The carbon loss and the nanocrystalline peak in the as-sprayed coating pattern indicate a more complex coating microstructure when compared to conventional coatings [2, 4]. The presence of these nanocrystalline phases in the as-sprayed coating was confirmed by the TEM analysis, which indicated approximately 300 nm-wide uniformly distributed particles. The d-spacing related to the different rings (Fig. 3b) and the theoretical d-saplings of different possible phases given in Table 2, indicate phases (W,WC,Co_3W_3C and Co), some of which were already detected with the single crystalline diffraction patterns (Fig. 3c, d). However the presence of other superimposed phases like W_2C, Co_6W_6C or $Co_3W_9C_4$ cannot be excluded. In general, the TEM analysis of as-sprayed coatings indicated a polycrystalline material with single crystal particles in the range of 100 and 300 nm.

Heating in an inert atmosphere above 600 °C leads to structural changes, bringing the phase composition closer to the equilibrium state. The $M_6C(Co_3W_3C)$ and $M_{12}C(Co_6W_6C)$, the so-called eta-phases, are possible equilibrium phases in the W-C-Co system, after loss of carbon in the spray process. Sometimes, their existence in the as-sprayed coatings is also reported, as summarized in earlier studies [2, 4]. After the heat treatment at 920 °C WC, metallic tungsten and two eta-phases $M_6C(Co_2W_4C$ fit the peaks better than those of $Co_3W_3C)$ and $M_{12}C$ (Co_6W_6C) were observed in the coating. This phase composition confirms the strong carbon loss. However, it should be mentioned that as a result of the heat treatment a decrease both in the carbon and oxygen contents is possible, due to the internal reduction processes. According to the W-C phase diagram, W_2C is thermodynamically unstable. Below 1250 °C it can decompose into WC and W during cooling of the WC-Co particle after impact. This decomposition can also result from the heat treatment at 920 °C. Annealing of the coating where Co is present will form the eta-phases as mentioned above, and Co can be fully consumed by these reactions. Hence W_2C was not observed in the XRD analysis (Fig. 2) of the HIPed coating. Similarly, the TEM analysis of the HIPed coating did not indicate the presence of Co (Table 3) due to its consumption in the formation of eta-phases.

The TEM analysis of the HIPed coating indicated approximately 20 nm-wide particles which were homogeneously distributed on the sample surface. The

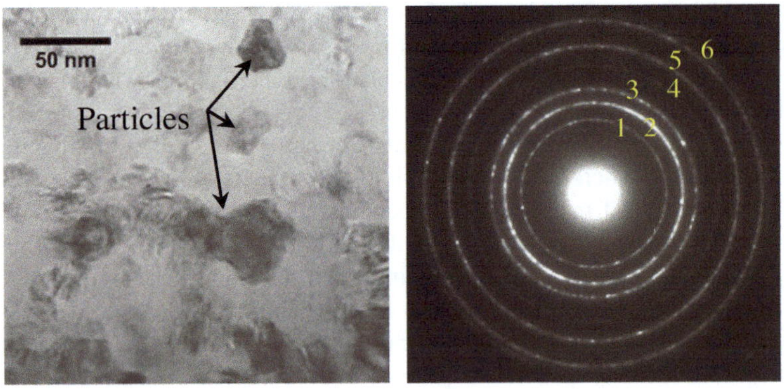

Fig. 4 TEM observations of the HIPed coating showing near 20 nm wide particles. The continuous lines (*rings*) in the diffraction pattern clearly indicate the nanocrystalline feature of the sample

Table 3 Experimental d-spacing and intensity measured from the different rings and theoretical d-spacing for some possible phases for the HIPed coating

		d_{th} (A)				
Ring number	d_{exp} (A)	WC (hex.)	W_2C (hex.)	Co_3W_3C or Co_6W_6C (cubic)	$Co_6W_9C_4$ (hex.)	W (hex.)
1	3 Int42	–	–	–	3.1 (201) Int7	–
2	2.4 Int100	2.5 (100) Int100	2.3 (−1−11) Int100	2.5 (331) Int40	2.6 (210) Int49	–
3	2.1 Int46	–	–	2.1 (511) Int100	2.2 (301) Int100	2.2 (011) Int100
4	1.7 Int16	–	1.7 (−1−12) Int15	1.7 (533) Int1	1.7 (401) Int11	–
5	1.5 Int35	1.4 (002) Int5	1.5 (300) Int16	1.5 (642) Int2	1.6 (214) Int20	1.6 (002) Int17
6	1.3 Int24	1.3 (111) Int15	1.3 (−1−13) Int15	1.3 (822) Int45	1.6 (305) Int55	1.3 (112) Int36

In brackets are the hkl values. The intensities of the X-ray reflections are also indicated

polycrystalline diffraction pattern from the sample surface is shown in Fig. 4. The d-spacing and intensities related to the different rings are given in Table 3. The high intensity of ring 2 is compatible with the presence of WC and/or W_2C on the sample surface. The observation of ring 1 indicates the presence of $Co_3W_9C_4$. The HIPing process therefore resulted in the crystallization of the amorphous phases along with the formation of eta-phases (M_6C,$M_{12}C$).

4.2 Sliding Wear

The sliding wear investigations of the as-sprayed and HIPed S-HVOF coating-ball couples (Fig. 5a) indicated that the total wear rate was on average lower for the HIPed coating. This is despite the higher averaged friction coefficient of the HIPed coating (Fig. 5b). The improvement in the total wear rate for the HIPed coating originated from a relatively lower ball wear rate when compared to the as-sprayed test couple. The lower ball wear rate could have also contributed to a higher friction coefficient for the HIPed test couples due to the relatively lower volume of ball-wear debris within the contact region. A transfer film consisting of oxidized Fe was observed in all test couples. These films were formed as a result of the flash temperature at asperity contact.

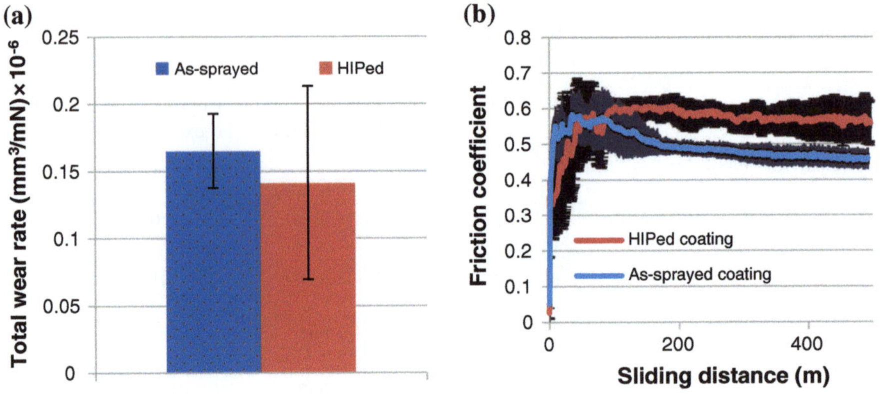

Fig. 5 Wear rate and friction comparison of the as-sprayed and HIPed coatings. **a** Total wear rate. **b** Friction coefficient

5 Conclusions

(1) The as-sprayed coating microstructure indicated 300 nm-wide uniformly distributed particles comprising of WC, W, Co_3W_3C and crystalline Co. A distinct nano-crystalline/amorphous phase was also detected in the as-sprayed coatings.
(2) Uniformly distributed 20 nm-wide particles were identified in the HIPed coating microstructure. HIPing transformed the coating microstructure bringing the phase composition closer to the equilibrium state. WC and $Co_3W_9C_4$ were detected on the sample surface. Co was not identified in the HIPed coatings.
(3) The sliding wear rate of the coating improved after the HIPing post-treatment whereas a reverse trend was observed for the friction coefficient.

References

1. P. Fauchais, G. Montavon, R.S. Lima, B.R. Marple, J. Phys. D: Appl. Phys. **44**, 53 pp, art no. 093001 (2011)
2. J. Oberste Berghaus, B. Marple, C. Moreau, J. Thermal Spray Technol. **15**, 676–681 (2006)
3. V. Stoica, R. Ahmed, T. Itsukaichi, S. Tobe, Wear **257**(11), 1103–1124 (2004)
4. R. Ahmed, N. H. Faisal, Nayef M. Al-Anazi, S. Al-Mutairi, F.-L. Toma, L.-M. Berger, A. Potthoff, E.K. Polychroniadis, M. Sall, D. Chaliampalias, M.F.A. Goosen, J. Thermal Spray Technol. **24**(3), 357–377 (2015)

Wavelength Depended Speckle Correlation Analyses of Engineered Metal Surfaces

Ersin Kayahan

Abstract Light source selection using rough surface illumination is crucial when especially high accuracy roughness measurements are required. It is therefore important to establish an extensive relationship between the wavelength of the illuminations, speckle size and roughness. Therefore, in this paper, an experimental approach for surface roughness evaluation based on the autocorrelation analysis of the spectral properties of speckle patterns caused by milled metal surfaces is reported. The speckle patterns obtained by the illumination of three 633, 604 and 543 nm wavelengths of He–Ne laser were also analyzed. It was found that wavelength depended speckle correlation analysis (WDSCA) is very sensitive to small variations in speckle sizes caused by spectral properties of speckle patterns such as increasing the wavelength lead to increased speckle sizes. The results are in good agreement with the results obtained from the mechanical stylus profilometer for the milled metal surfaces with roughness values between R_a; 0.36 and 1.98 μm. The technique reported here has a great potential for precise and non-contact optical measurements of rough surfaces.

1 Introduction

In recent decades, surface parameters have become important variables in various industrial processes or production techniques. The surface roughness parameters are important during manufacturing to ensure the satisfactory performance

E. Kayahan (✉)
Electro-Optic and System Engineering, Kocaeli University, 41380 Umuttepe, Kocaeli, Turkey
e-mail: kayahan@kocaeli.edu.tr

E. Kayahan
Kocaeli University-LATARUM Laboratory, 41275 Yenikoy, Kocaeli, Turkey

E. Kayahan
Barbaros Denizcilik YO, Kocaeli University, 41500 Karamürsel, Kocaeli, Turkey

E.K. Polychroniadis et al. (eds.), *2nd International Multidisciplinary Microscopy and Microanalysis Congress*, Springer Proceedings in Physics 164,
DOI 10.1007/978-3-319-16919-4_6

characteristics of the work-piece and concomitant tool wear [1]. The roughness is defined as the irregularities on a material surface as a result of various machining operations. The symbol R_a is commonly used to describe it in literature as average roughness. Theoretically, R_a is the arithmetic average value of departure of the profile from the mean line throughout the sampling length. R_a is also an important factor in controlling machining performance during end milling operations [2]. Nevertheless, the surface roughness measurements based on the stylus instrument have some drawbacks such as possible surface damages, long measuring times and difficulty of in situ measurement process. These drawbacks have prompted the development of alternative techniques including optical methods such as optical interference and light scattering techniques for the surface roughness measurement as non-contact optical methods leading to non-destructive, fast and continuous measurements [2]. With the technological progress in the field of optoelectronics (lasers, CCD and CMOS-cameras), optical metrology received considerable attention to measure the surface properties as non-contact testing. A number of studies have been reported in literature evaluating the surface roughness on rough surfaces by using non-contact optical methods such as the mean scattered intensity (MSI), laser interferometry and speckle pattern analysis methods [3–7].

On projecting a coherent light to rough surface, the reflected waves from different points of the surface interfere and form speckles pattern. The pattern is due to the destructive interference as the phase difference between waves is equal to half of the wave length of illumination light used ($\lambda/2$). A surface having roughness of $\lambda/4$ gives a path difference of $\lambda/2$ for normal incidence. This means is the obtained speckle pattern depending on the wave length of illumination light and the pattern carries important information about the roughness of the surface. An experimental approach has been introduced in [7] based on the speckle pattern correlation technique for the measurement of the surface roughness on the machined surfaces such as grinding and milling. The speckle patterns of the machined surfaces are captured using a collimated laser beam (He–Ne laser, 10 mW, $\lambda = 633$ nm) and a CCD camera. A theoretical approach depending on the speckle pattern has been presented for the measurement of the surface roughness in [4]. Toh et al. [8] reported a speckle correlation technique for the estimation of the surface roughness ranging from 1.6 to 50 μm at different incidence angles (5°, 15°, 30°, 45°) in the far-field plane. Persson [5] has presented an experimental work evaluating the surface roughness by utilizing angular speckle correlation (ASC) technique on the machined surfaces where a laser beam (He–Ne) is employed to illuminate the surface at an incidence angle of 45°. Persson [9] also presented a speckle contrast technique for real-time roughness measurements. He used only a CCD camera and a laser in his study and he achieved measurements of roughness in the range of R_a (using stylus instruments) 0.010–0.150 μm using the speckle contrast technique. Usually speckle contrast methods can evaluate surface roughness values less than $R_a < 0.3$ μm, while the speckle correlation methods can work anywhere between 1 and 30 μm.

Wong and Li [10] focused on a new optical system which utilizes the combined effects of interference and light scattering for the surface roughness measurements in a grinding process. Tay et al. [3] have proposed a light scattering technique to measure the roughness of semi-conductor wafers. This technique utilizes a low power He–Ne laser beam and records surface roughness in the nanometer range with a high degree of accuracy. Tian et al. [11] showed that a new approach integrating speckle pattern correlation and light scattering methods to evaluate surface roughness for a measuring range from 0.05 to 1.6 μm. This approach employs a laser and two cameras. The laser emits a certain light pattern on the surface to be measured and while camera 1 is used to measure the surface form and waviness using the light scattering method, camera 2 is used to control surface roughness with speckle pattern. Lehman and Goch [12] developed new speckle techniques including the method of poli-chromatic speckle autocorrelation and the doubly scattered light approach for moving metal surfaces.

The aim of this study is to find correlation between illumination wavelengths with speckle size and to determine suitable illumination wavelength for the investigated roughness values of the metal surfaces. So, in this paper, the spectral properties of speckle patterns of face-milled metal surfaces were analyzed by means of statistical autocorrelation method obtained thorough three wavelengths of CW He–Ne laser illuminations (λ = 633, 604 and 543 nm). Firstly, speckle patterns of the surfaces are captured using each wavelength of He–Ne laser and CCD camera containing 1024 × 1360 pixels. Then, the variation of various autocorrelation parameters of speckle images and surface roughness (stylus) were investigated for each illumination wavelength. It was found that as the wavelengths increases, so does the average speckle size. The results are in good agreement with the surface roughness values obtained by the mechanical system (the stylus method).

2 Theory

When a surface is illuminated by a light wave, each point on an illuminated surface acts as a source of secondary spherical waves. The light at any point in the scattered light field is made up of waves which have been scattered from each point on the illuminated surface. If the surface is rough enough to create path-length differences exceeding one wavelength, giving rise to phase changes greater than 2π, the amplitude, and hence the intensity of the resultant light varies randomly. So, a grainy pattern caused by the interference of the laser light which is irradiated to rough surface by the random phase based on the surface roughness is observed in the reflected beam. Smooth surfaces scatter laser light mainly in specular direction. When the surface got rougher, the specular component decreased and the scattering component was increasing (Fig. 1a). For an optically smooth surface with the heights of surface irregularities much smaller than wavelength of the incident light, the relation between the scattered light and the surface roughness has been explained by use of vector diffraction and Bechmann scalar theory [7, 13]. So, it is shown by some researchers that scattered light intensity distribution is strongly depended on surface roughness and this used as non-contact roughness measurements [9, 11].

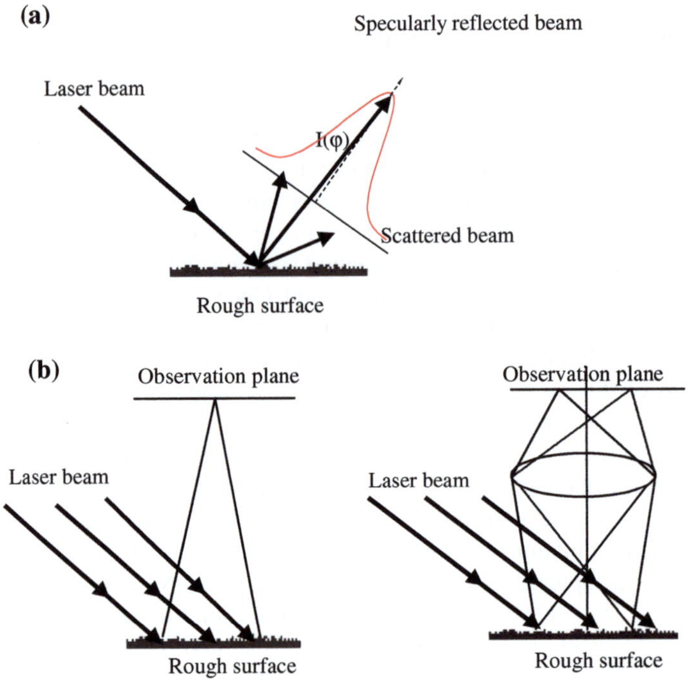

Fig. 1 Scattered laser beam from a rough surface and its corresponding light distribution (**a**), two basic schemes of speckle pattern formation (**b**)

The two basic types of the speckle are specified as objective and subjective as seen in Fig. 1b. When the speckle is recorded directly by means of any 2D detector without any imaging optics in the free space, the obtained pattern is called "objective" speckle. The average size of the speckles in these patterns is determined by the wavelength of the laser, the size of illuminated surface area and the geometry involved. If the object distance to detecting surface is L and illuminated area has diameter D, the wavelength of the laser beam is λ, then the average diameter of the objective speckle is;

$$d = 1.2\lambda \frac{L}{D} \tag{1}$$

It is clear that for a given geometry (L/D) = const., the average speckle size is proportional to the incident laser wavelength.

Another type of speckles is called "subjective" speckles. Every imaging system (human eye, microscope objective, etc.) between reflecting surface and recording detector alters the coherent superposition of waves and thus produces its own

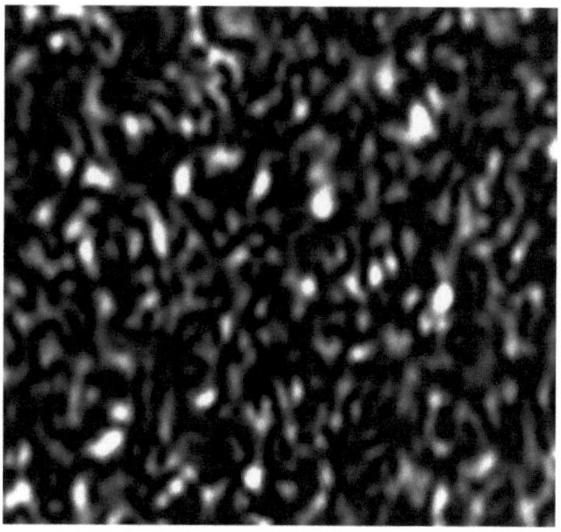

Fig. 2 Example of the speckle pattern obtained using the optical setup seen in Fig. 3

speckle size. The detailed structure of the speckle pattern depends on the viewing system parameters; for instance, if the size of the lens aperture changes, the size of the speckles change, too. If the position of the imaging system is altered, the pattern will gradually change and will eventually be unrelated to the original speckle pattern. Detailed information on "subjective" speckles can be found elsewhere [14]. Final formula for the average diameter of the subjective speckle is;

$$d = 1.2\lambda(1 + M)\frac{f}{D} \tag{2}$$

where, M is the imaging system magnification, f is focal distance and λ is the wavelength of the laser beam. It is obvious that for a given surface and constant imaging system parameters, the average speckle size will increase with increasing wavelength.

The speckle patterns obtained in our experiments are the "subjective" speckles. Figure 2 depicts a typical pattern obtained from our experimental set-up with a roughness of $R_a = 0.770$ μm.

3 Experimental

Our experimental studies can be separated into two stages: first one is the milling and measuring of mechanical roughness of the specimens and the second is the optical system to obtain the speckle pattern. The optical system shown schematically in Fig. 3 was designed for obtaining speckle patterns of the specimens. Three wavelengths of CW He–Ne laser (λ = 633, 604 and 543 nm) were used to illuminate

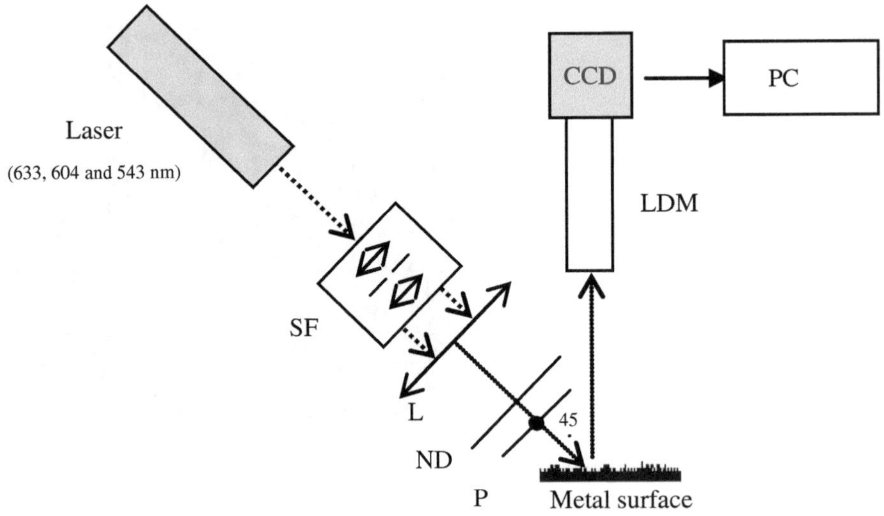

Fig. 3 Schematic diagram for speckle pattern formation where LDM, L, ND, P and SF are long distance microscope, lens, neutral density filter, pinhole and spatial filter, respectively

the sample surfaces where the angle of illumination is 45°. A spatial filter (f/4) was utilized to create smooth laser transverse intensity profile. A lens (f = 50 mm) was employed to produce a small intense light spot on the specimen surface. The scattered light was recorded by a long distance microscope (LDM) (8x) and by CCD camera Prosilica-GC11350 with a full resolution of 1024 × 1360 pixels, placed perpendicularly to the surface.

The output power of the He–Ne laser at various wavelengths is different. On other hand, the spectral sensitivity of CCD pixels also varies with wavelength. We have therefore performed a radiometric calibration of optical system using a smooth white surface. The output power of laser at 543 nm (lowest for this laser) was attenuated by a set of neutral density filters to a desired value at which the gray level of the reflected beam from the reference surface image pattern at a given exposure time was below saturation. The grey level histogram of this image was selected as a reference for the intensity calibration of other (λ = 633 and 604 nm) wavelengths, whose histograms were adjusted by means of additional neutral density filters and small changes in exposure times. The beam was plane polarized at every wavelength.

Distance between the metal surface (L) and LDM was kept constant for all the surface measurements. The illuminated area of the surface was 1 mm^2. Speckle pattern images were recorded for the metal surfaces of roughness values R_a; 0.36–1.98 μm.

4 Results

In Fig. 4. shows the digital speckle patterns for 543, 603 and 633 nm illumination wavelengths. A correlation analysis together with computation of several parameters which established a quantitative relationship between speckle pattern and surface roughness has been introduced in [15]. As a figure of merit, various parameters of the normalized autocorrelation (AC) function were employed as given below;

$$AC = \frac{\sum_{i=1}^{M} \sum_{j=1}^{N} (g(i,j) - g_0)(g(i + \delta i, j + \delta j) - g_0)}{\sum_{i=1}^{M} \sum_{j=1}^{N} g^2(i,j)} \tag{3}$$

Here, $g(i,j)$ is the actual grey level obtained from subtraction of noise signal obtained from the dark image of the CCD camera from the recorded original grey level and g_0 is the mean intensity value. The integers, $\delta_i = 1, \ldots, M$ and $\delta_j = 1, \ldots, N$, are pixel shift values. In this study, we examined the observed patterns with the following parameters:

(1) Integrated autocorrelation coefficient, A*CI (X),* (shift in x-direction of lag length δj which varies between 1 and N)

$$ACI(X) = \sum_{i=1}^{M} \frac{\sum_{i=1}^{M} \sum_{j=0}^{N-\delta j} g(i,j)\, x\, g(i, j + \delta j)}{\sum_{i=1}^{M} \sum_{j=1}^{N} g^2(x,y)} \tag{4}$$

(2) Integrated autocorrelation coefficient ACI *(Y)* is calculated shifting the image in the y-direction with steps of δi which varies between 1 and M.

$$ACI(Y) = \sum_{i=1}^{M} \frac{\sum_{i=1}^{M-\delta i} \sum_{j=1}^{N} g(i,j)\, x\, g(i + \delta i, j)}{\sum_{i=1}^{M} \sum_{j=1}^{N} g^2(x,y)} \tag{5}$$

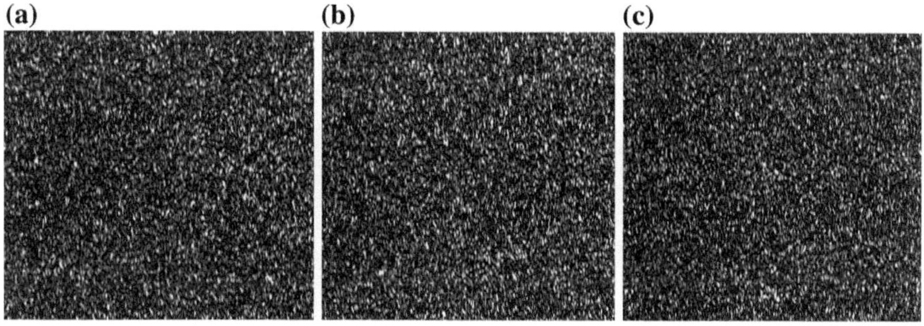

Fig. 4 CCD image (1024 × 1360 pixels) of a speckle pattern obtained from the surface of a milled specimen with roughness of (R_a = 0.770 μm). The pattern obtained from the same surface with different illumination wave lengths, 633 nm (**a**), 604 nm (**b**) and 543 nm (**c**)

(3) Integrated autocorrelation coefficient A*CI (XY) is* calculated shifting the image in the diagonal direction with steps of δi and δj where i and j vary between 1 and M, 1 and N, respectively.

$$ACI(XY) = \sum_{i=1}^{M} \frac{\sum_{i=1}^{M-\delta i} \sum_{j=1}^{N-\delta j} g(i,j)\, x\, g(i+\delta i, j+\delta j)}{\sum_{i=1}^{M} \sum_{j=1}^{N} g^2(x,y)} \tag{6}$$

Figure 5 shows the variation of ACI(X), ACI(Y), ACI(XY) with laser wavelength for the metal surface with R_a; 0.36 μm. Since we have used normalized autocorrelation functions, their values are equal to unity at the origin. By increasing the shift in both directions the function decreases to a certain constant value. This behavior was observed for all patterns. The fall speed autocorrelation functions characterize the speckle size. The decrease of the fall speed with increasing wavelength was observed for all patterns. Table 1 shows integrated ACI(X), ACI (Y), ACI(XY) autocorrelation function parameters versus illumination wave

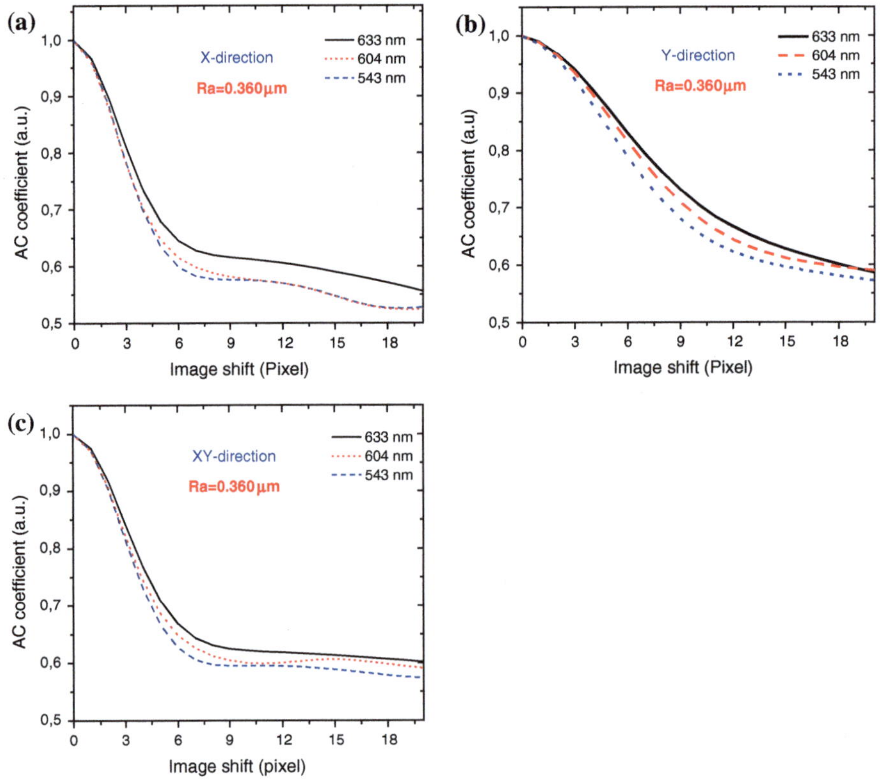

Fig. 5 Changes of AC function of speckle patterns with laser wavelength for 0.360 μm rough surface. Shifting was carried out in X, Y and diagonal (XY) directions (**a**), (**b**) and (**c**), respectively

Table 1 Variation of integrated autocorrelation function parameters with illumination wavelengths and mechanical roughness (Stylus) values of the surfaces

Sample no.	Stylus values	633 nm illumination			604 nm illumination			543 nm illumination		
		Int. autocorrelation coefficient			Int. autocorrelation coefficient			Int. autocorrelation coefficient		
	Ra (μm)	IAC-X	IAC-Y	IAC-XY	IAC-X	IAC-Y	IAC-XY	IAC-X	IAC-Y	IAC-XY
1	0.360	156.2174	152.1176	134.5145	138.3868	137.1202	121.2212	133.5649	136.9829	121.0638
2	0.413	156.2277	155.2394	134.7013	139.3549	138.351	121.6672	136.704	141.4245	121.6494
3	0.450	157.199	156.6863	135.6276	143.3062	139.3347	126.0777	144.8737	143.2181	128.4722
4	0.670	159.2156	159.3406	138.2788	148.2102	148.9965	134.0015	149.0595	148.0838	132.1933
5	0.706	160.6678	159.8332	138.4112	149.1562	152.0403	134.0244	149.7436	148.4158	132.4543
6	0.750	161.6911	160.6005	139.5398	149.1842	153.2824	134.4758	150.0532	148.6736	134.0807
7	0.823	163.7341	164.6321	142.6069	150.5154	154.7441	137.2265	152.5201	153.7396	136.033
8	0.956	164.0149	165.4487	143.5762	150.7981	155.116	140.0351	154.349	154.4959	136.106
9	1.690	164.607	166.0913	145.6876	155.5794	155.3813	140.7755	156.0183	154.9046	137.1286
10	1.746	165.0808	166.774	147.2061	157.2472	156.7165	142.9107	156.2261	155.0698	137.1345
11	1.980	166.2195	168.727	147.4786	160.3411	159.999	146.8127	157.9691	160.8851	137.3757

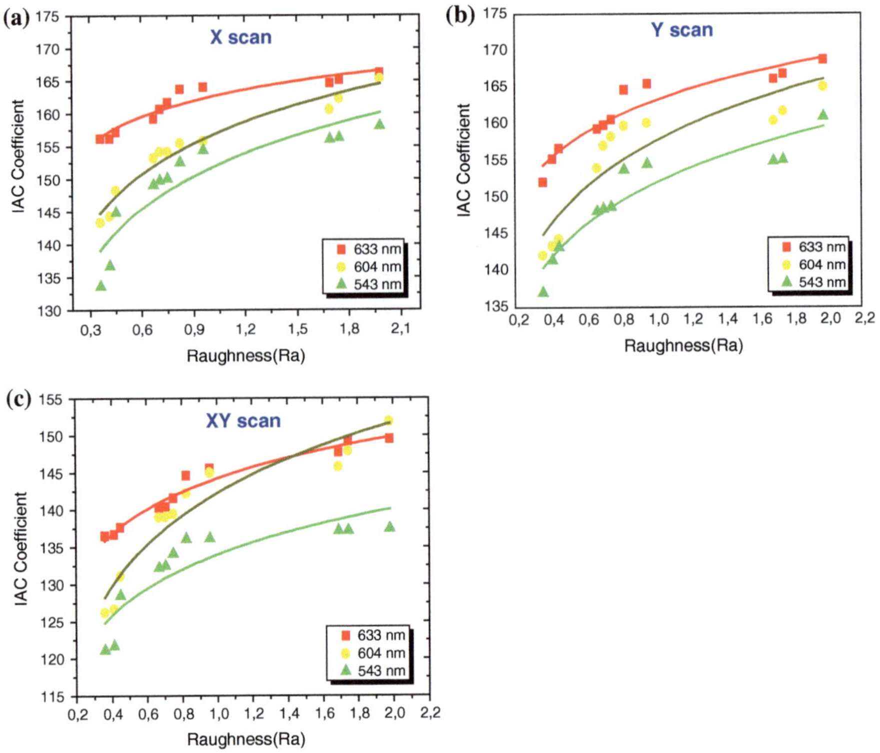

Fig. 6 Variation stylus roughness of the specimens with integrated autocorrelation (IAC) function for the various shifting scans of X scan (**a**), Y scan (**b**) and XY scan (**c**). *Solid lines* show corresponded fit function

Table 2 The fitted function parameters with illuminated the laser wavelengths

Selected fit function (IAC = A * ln(Ra) + B) Fit parameters				
Illumination (nm)	Shifting	A	B	R^2
633	X	15.930 ± 0.659	156.571 ± 0.393	0.89986
	Y	8.621 ± 0.923	154.638 ± 0.551	0.90652
	XY	8.111 ± 0.556	136.015 ± 0.332	0.96628
604	X	11.562 ± 0.778	144.697 ± 0.464	0.96082
	Y	12.430 ± 1.935	145.077 ± 1.155	0.82092
	XY	13.70 ± 1.395	128.193 ± 0.833	0.9145
543	X	12.272 ± 1.843	139.154 ± 1.100	0.83147
	Y	11.231 ± 1.365	140.323 ± 0.815	0.88284
	XY	8.895 ± 1.631	124.819 ± 0.974	0.76752

lengths. It has also been observed that the parameters decrease with decreasing illumination wavelengths.

Figure 6 shows the variation of IACX, IACY and IACXY obtained different illuminations with surface roughness for the surfaces. Where, a good correlation is observed between IAC coefficient with the roughness values. It can be shown that IACX, IACY and IACXY values increase logarithmically with the roughness values. It would be understand by wavelength dependent speckle pattern. An empirical equation can be used as fitting function. The fitting parameters is listed in Table 2 for each the illuminations. The highest correlation is obtained from XY shifting for 633 nm illumination (R^2 = 0.96628). It can be therefore conclude that the using of illumination of 633 nm as light source would be more suitable to determine roughness of metal surfaces for the given roughness values.

5 Conclusions

An experimental investigation of wavelength depended speckle correlation analysis (WDSCA) for metal surfaces was studied. WDSCA is very sensitive to small variations in speckle sizes caused by spectral properties of speckle patterns such as increasing the wavelength lead to increased speckle sizes. IAC coefficients are good agreement with roughness values of the surfaces and they can be determined a logarithmic function. Light source selection using surface illumination is crucial when especially high accuracy roughness measurements are required. It is therefore important to establish an extensive relationship between the wavelength, speckle size and roughness. Illumination with light source of 633 nm has more suitable to determine roughness of metal surfaces for the given roughness values for this study. WDSCA technique could be great potential for non-contact optical roughness measurements of various surfaces at near future.

Acknowledgments This research was financially supported by the Scientific Research Projects Unit of Kocaeli University (Project no. 2007/29) and Scientific and Technological Research Council of Turkey (TUBITAK) under grant no 209T123. The author would like to thanks TSE mechanics laboratory frameworks for helping mechanical roughness measurements and Dr. H. Oktem for his help in sample preparation and Prof. O. Gundogdu, Dr. H. Nasibli and Prof. F. Hacizade for helpful discussions.

References

1. M.F. Ruiz Gale, E.N. Hogert, N.G. Gaggioli, Opt. Laser Technol. **45**, 947(2007)
2. M. Sander, *A Practical Guide to the Assessment of Surface Texture*, *Germany* (1991)
3. C.J. Tay, S.H. Wang, C. Quan, C.K. Ng, Surface roughness measurement of semi-conductor wafers using a modified total integrated scattering model. Optic **113**, 317 (2002)
4. B. Ruffung, Application of speckle-correlation methods to surface roughness measurement: a theoretical study. J. Opt. Soc. Amer. A **3**, 1297 (1986)

5. U. Persson, Surface roughness measurement on machined surfaces using angular speckle correlation. J. Mater. Process. Technol. **180**, 233 (2006)
6. D.J. Whitehause, *Handbook of Surface and Nanometrology* (Institute of Physics Publishing (IOP), London, 2003)
7. B. Dhanasekar, N.K. Mohan, B. Bhaduri, B. Ramamorthy, Evaluation of surface roughness based on monochromatic speckle correlation using image processing. Precision Eng. **32**, 196 (2008)
8. S.L. Toh, C. Quan, K.C. Woo, C.J. Tay, H.M. Shang, Whole field surface roughness measurement by laser speckle correlation technique. Opt. Laser Technol. **33**, 427 (2001)
9. Persson, U., Real time measurement of surface roughness on ground surfaces using speckle-contrast technique. Opt. Laser Eng. **17**(2), 61 (1992)
10. P.L. Wong, K.Y. Li, In-process roughness measurement on moving surfaces. Opt. Laser Technol. **31**, 543 (1999)
11. G.Y. Tian, R.S. Lu, D. Gledhill, Surface measurement using active vision and light scattering. Opt. Laser Eng. **45**, 131 (2007)
12. P. Lehmann, G. Goach, Comparison of conventional light scattering and speckle techniques concerning an in-process characterization of engineered surfaces. Ann. CIRP **49**(1), 419 (2000)
13. J.M. Elson, J.M. Bennett, Vector scattering theory. Opt. Eng. **18**, 116 (1979)
14. W. Lauterborn, T. Kurz, *Coherent Optics: Fundamentals and Applications*, 2nd edn. (Springer, Berlin, 2002)
15. S.L. Toh, H.M. Shang, C.J. Tay, Surface-roughness study using laser speckle method. Opt. Laser Technol. **29**, 217 (1998)

Effect of V_2O_5 Additives to the Sintering of Y_2O_3

Tarik Talib Issa, Kawakib Jassim Majeed and Jenan M. Hasan

Abstract The effect of sintering additive on the microstructure development of yttrium oxide was investigated at different sintering temperature (700, 900, 1100, 1300 °C) under static air. Different combinations of yttrium oxide with V_2O_5 were used as a ceramic composite material. X-ray diffraction, differential thermal analysis (DTA) were investigated and microstructure of resulting compacts were characterized by, using scanning electron microscopy, porosity and sintering density reflected the optimum values for the combination Y_2O_3–3 wt% V_2O_5 sintered at temperature 1300 °C. Mechanical properties representative by hardness and compression were tested. Brake down voltage and dielectric strength were measured for all compacted samples sintered at 1300 °C.

1 Introduction

Yttria has received great attention in various fields of advanced applications due to its excellent properties such us high dielectric constant(14–18), low absorption in broad range (near—UV to IR), superior electrical brake—down (>3 MV/cm) and low leakage current [1]. The remarkable luminescence render it widely be used in fluorescent lighting, color television, computer monitors, flat panel display, X-ray imaging and amplifiers for fiber—optics communication [2, 3]. Its is a refractory material using in coating crucibles, tubers and nozzles for improved resistance to molten titanium and metals. As ceramic it is a super-excellent media for high brightness laser and Yttria—stabilized zirconia was used in alumina—zirconia

T.T. Issa (✉) · K.J. Majeed · J.M. Hasan
College of Science, Physics Department, University of Baghdad, Jadiriya, Baghdad, Iraq
e-mail: alazbrh@yahoo.com

K.J. Majeed
e-mail: alazbrh@yahoo.com

J.M. Hasan
e-mail: alazbrh@yahoo.com

E.K. Polychroniadis et al. (eds.), *2nd International Multidisciplinary Microscopy and Microanalysis Congress*, Springer Proceedings in Physics 164,
DOI 10.1007/978-3-319-16919-4_7

abrasives, bearings and seals, high temperature refractories for continuous—casting nozzles, jet-engine coating, oxygen sensors in automobile engines, and wear-resistant and corrosion—resistant cutting tools [4]. Sintering is a manufacturing process in which a fine powder that has been formed into a shape is subsequently fired at high temperature. The compact, then fired, densifies and becomes non-porous. More formally, Sintering is a thermal treatment that bonds particles together into a solid, coherent structure by means of mass transport mechanisms occurring largely at the atomic level A [5]. Variety of new ceramics has been developed in the last 20 years. These are of particular interest because they have either unique or outstanding properties and greater chemical receptivity or they have been discovered more or less accidentally and have become an important part of the industry Because there is a real need for new materials to transform presently available designs into practical serviceable products. By far the major hindrance to the development of many new technologically feasible structures and systems is the lack of satisfactory materials. Advanced re constantly filling this need. Other new ceramic materials unknown 10 or 20 years ago are no being manufactured. From this point of view the ceramic industry is one of our most rapidly changing industries with new products having useful properties constantly being developed, these ceramics [6]. In the present work we are focusing on improvement in the microstructure of Yettria by adding additives oxide like vanadium penta oxide, which showed a great effecting at different Sintering temperature comparing with other oxide [7].

2 Experimental

The Y_2O_3 and V_2O_5 powder, was analyzed for particle sizes ranging between (50–70) microns were generally utilized as starting materials throughout the present investigation. V_2O_5 at varying fractions (1, 2, and 3 wt%) was dry mixed with Y_2O_3 by using mixing technique. Discs of 2 cm diameter were compacted, three samples for each weight percentage were prepared. Sintering was performed at a various temperatures ranging from (700, 900, 1100, 1300 °C) for 2 h under static air. X-ray diffraction was carried out by SHEMADU XRD—600 (Japan), and differential thermal analysis, DTA by using LINSEIS STA (Germany), for the combination Y_2O_3—3 wt% V_2O_5 were done, DTA was done at 5 °C/min. and from (25–1000 °C), in an ambit atmosphere. Densification measurements were based on volumes determined using micrometer measurements and accurate weight measured by an electronic balance. Microstructure examination was carried out for the sintered specimens using scanning electron microscope SEM—VEGA, TESCAN (Geake Republic). Hardness and compression strength tested to. Finally brake down voltage and dielectric strength were tested for the all combination sintered at 1300 °C, under static air.

3 Results and Discussion

Figure 1, shows the X-ray diffraction pattern for the sintered combination Y_2O_3—wt% V_2O_5 sintered at 1300 °C, its very clear form its profile that there is no any phase changes after adding V_2O_5 as Sintering aid and the resulting spectrum is belongs to Yttria. [8] Thermal analysis (DTA), gives a consistency results with X-ray diffraction, by no phase transformation (Exo.) [8], can be notes after the sintering and adding vanadium pent oxide, while decomposition notes at 65 °C, which is belong to the water molecules caused by the surrounding atmosphere as shown in Fig. 2. In Fig. 3, we can observe the sintering density of the Y_2O_3(1, 2 and 3 wt% V_2O_5, sintered at various sintering temperature under static air for 2 h. All The compacts containing wt% V_2O_5 showed an increasing in density with maximum value reached at (2.92 gm/cm^3) for the combination Y_2O_3–V_2O_5 3 wt% sintered at 1300 °C. Sintering process also effecting the porosity to be reduced for Y_2O_3 and Y_2O_3—wt% V_2O_5 respectively from (8.40) and to be at a lowest value (7.08) for the Y_2O_3—3 wt% V_2O_5 sintered at 1300 °C, which giving the consistency with the sintering behavior [8], as shown in Fig. 4. The micro structure investigation of the ceramic compacts showed results consistent with the sintered

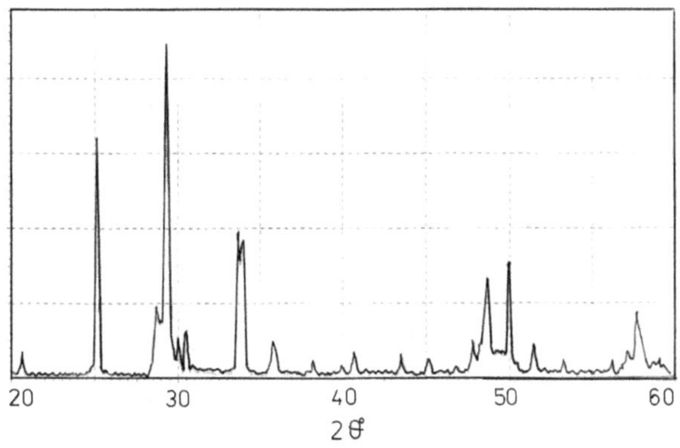

Fig. 1 The X-ray diffraction pattern of Y_2O_3—wt% V_2O_5 ceramic compacts, sintered at 1300 °C, for 2 h under static air

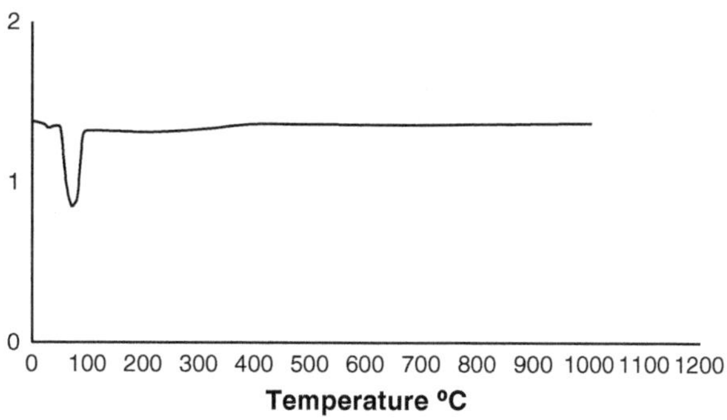

Fig. 2 The DTA curve of Y_2O_3—wt% V_2O_5 ceramic compacts, sintered at 1300 °C for 2 h under static air

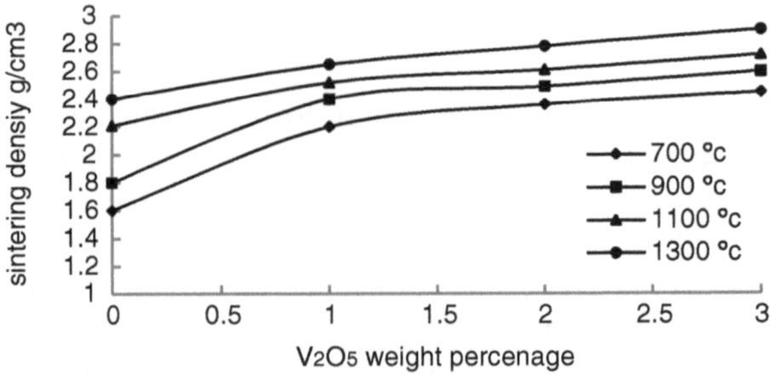

Fig. 3 The sintered density of Y_2O_3—wt% V_2O_5 ceramic compacts, sintered at different sintering temperature for 2 h under static air

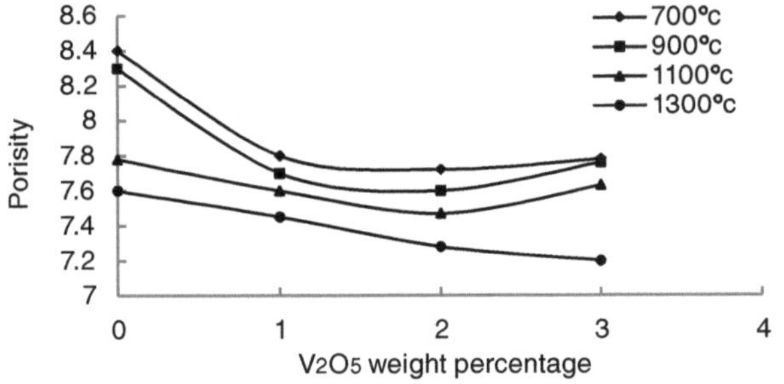

Fig. 4 The porosity of Y_2O_3—wt% V_2O_5 ceramic compacts, sintered at different sintering temperature for 2 h under static air

densities. Some representative SEM micrographs for samples sintered at 1300 °C, for 2 h in air are shown in Fig. 5. The compact containing 1 wt% V_2O_5 showed evidence of the fine 50 micron of grain size, indicating the first sintering stage Fig. 5a. Whereas the compact containing 2 wt% V_2O_5 showed the intermediate

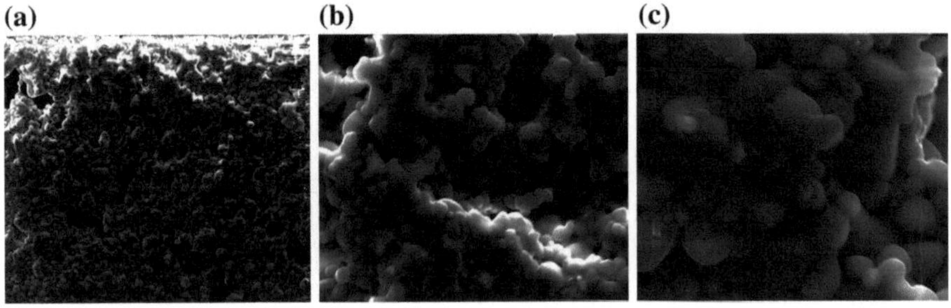

Fig. 5 SEM micrographs for **a** Y_2O_3–1 wt% V_2O_5 of 50 mμ grain size, **b** Y_2O_3—2 wt% V_2O_5 10 mμ grain size, **c** Y_2O_3—3 wt% V_2O_5, 5 mμ grain size. Sintered at 1300 °C for 2 h under static

Table 1 The vickers hardness and compression strength for the ceramic composite Y_2O_3—wt% V_2O_5 sintered at 1300 °C, for 2 h under static air

Ceramic compacts	Vickers hardness (MPa)	Compression (MPa)
Y_2O_3	19.57	1.40
Y_2O_3–V_2O_5 1 wt%	23.91	1.30
Y_2O_3–V_2O_5 2 wt%	28.97	10.47
Y_2O_3–V_2O_5 3 wt%	30.21	12.31

Table 2 The brake down voltage and the dielectric strength for the ceramic composite Y_2O_3—wt % V_2O_5 sintered at 1300 °C, for 3 h under static air

Ceramic compacts	Brake down voltage (kv)	Dielectric strength (kv/mm)
Y_2O_3	12.43	2.55
Y_2O_3–V_2O_5 1 wt%	14.40	2.53
Y_2O_3–V_2O_5 2 wt%	16.50	3.09
Y_2O_3–V_2O_5 3 wt%	22.50	4.59

sintering stage of 10 micron coarse particle size, begins when adjacent necks begin to impinge upon each other, densification and grain growth occur during this stage with open–pore structure [9, 10], Fig. 5b, comparing with the 3 wt% V_2O_5 as shown in Fig. 5c, the grain growth of 5 micron with closed porosity indicate the presence of the final stage of sintering caused by V_2O_5 particles that improve the Sintering activity of Y_2O_3 powder through the bulk transport mechanism result in net particle movement leading to the densification [7, 11, 12].Which shows a significant increase in both hardness and compression strength especially at the combination Y_2O_3—3 wt% V_2O_5, as shown in Table 1. And the same effect can be identify for the brake down voltage and dialectical strength having a maximum value of 22.5 (Kv), and 4.59 (Kv/mm), respectively, as shown in Table 2.

4 Conclusions

The densification of the Y_2O_3–V_2O_5 wt% ceramics compacts increased at all sintering temperatures and indicate a higher sintering density at 1300 °C, for the combination Y_2O_3–V_2O_5 3 wt% shows the final Sintering stage with a grain growth of 5 microns and close pore. Vickers hardness and compression strength shows the same behavior like in sintering density, meanwhile both brake down voltage and dielectric constant improved cumbering it with the matrix. TiO_2 was found to improve the sintering activity, mechanical and electrical properties of Y_2O_3 and without any effecting to its crustal structure.

Acknowledgments The externally grateful to the all people that support, and facilities offered throughout this work, specially to the material labs staff.

References

1. L. Manchanda, M. Gurvitch, I.E.E.E. Electron, Dev. Lett. **51**, 919 (1988)
2. M.J. Weber, J. Lumin. **100**, 35–45 (2002)
3. D.W. Cooke, B.L. Bennett, R.E. Muenchausen et al., J. Lumin. **106**, 125–132 (2004)
4. R.M. German, *Sintering Theory and Practice* (Wiley, New York, 1996)
5. G.T. Nolan, P.E. Kavanagh, The size distribution of interstices in random pickings of spheres. Powder Technol. **78**, 231–238 (1994)
6. D.A. Taylor, Materials Australia **33**(1):20–22 (2001)
7. T.T. Issa, K. Jassim, H.J. Ibrahim, Effecting of TiO2 additives to the sintering of Y2O3. American Institute of Physics AIP Conference Proceedings, pp. 1569, 299 (2013)
8. G.-X. Liu, G.-Y. Hong, J.-X. Wang, Hydrothermal synthesis of spherical Y2O3: Eu Luminescent material with smooth surface. Chin. J. Inorg. Chem. **22**(8), 1535–1539 (2006)
9. W.S. Slaughter, et al., A quantitative analysis of the effect of geometric assumption in sintering models. Acta Metallurgica **54**(12), P 5077 (1997)
10. A. Bellis, Computer modeling of sintering in ceramics, M.Sc. thesis, Graduate Faculty School of Engineering, University of Pittsburgh, June 20 2004
11. T.T. Issa, Sintering behavior of MgO ceramic compacts with and without dopants. In: The 3rd Scientific Conference of the College of Science, University of Baghdad, p. P2022, 24 March 2009
12. J. Tong, D. Clark, L. Bernau, M. Sanders, O'. Rayan, Grained yttrium—doped barium zirconate proton conduction ceramics. J. Mater. Chem. **20**, 6333 (2010)
13. T. Andelman, S. Gordonov, G. Busto, P.V. Moghe, R.E. Riman, Synthesis and cytotoxicity, of Y2O3 nanoparticles of various morphologies. Nano Scale Res. Lett. **5**, 263–273 (2010)

Integrating Microscopic Analysis into Existing Quality Assurance Processes

Peter Frühberger, Thomas Stephan and Jürgen Beyerer

Abstract When technical goods, like mainboards and other electronic components, are produced, quality assurance (QA) is very important. To achieve this goal, different optical microscopes can be used to analyze a variety of specimen to gain comprehensive information by combining the acquired sensor data. In many industrial processes, cameras are used to examine these technical goods. Those cameras can analyze complete boards at once and offer a high level of accuracy when used for completeness checks. When small defects, e.g. soldered points, need to be examined in detail, those wide area cameras are limited. Microscopes with large magnification need to be used to analyze those critical areas. But microscopes alone cannot fulfill this task within a limited time schedule, because microscopic analysis of complete motherboards of a certain size is time demanding. Microscopes are limited concerning their depth of field and depth of focus, which is why additional components like XY moving tables need to be used to examine the complete surface. Yet today's industrial production quality standards require a 100 % control of the soldered components within a given time schedule. This level of quality, while keeping inspection time low, can only be achieved when combining multiple inspection devices in an optimized manner. This paper presents results and methods of combining industrial cameras with microscopy instrumenting a classificatory based approach intending to keep already deployed QA processes in place but extending them with the purpose of increasing the quality level of the produced technical goods while maintaining high throughput.

P. Frühberger (✉) · T. Stephan · J. Beyerer
Fraunhofer Institute of Optronics, System Technologies and Image Exploitation IOSB, Fraunhoferstr. 1, 76131 Karlsruhe, Germany
e-mail: peter.fruehberger@iosb.fraunhofer.de

E.K. Polychroniadis et al. (eds.), *2nd International Multidisciplinary Microscopy and Microanalysis Congress*, Springer Proceedings in Physics 164,
DOI 10.1007/978-3-319-16919-4_8

1 Introduction

Quality Assurance (QA) plays an important role in nowadays industrial production. The solution of choice are industrial cameras, which are an established technique and are widely used in order to do completeness checks. Industrial cameras are quite cost-efficient and can be combined with different lenses to meet the requirements of the inspection task. An overview of different inspection methods instrumenting industrial cameras is given in [1].

There are use-cases, for example inspection of SMD-Solder points, where it is not constructive to build microscopy-like inspection systems out of custom equipment just to fulfill the very same task a microscope could handle. Cost efficiency is getting worse and market-ready systems are already available that can be integrated directly in the current QA process. In fact most of the time randomly chosen samples out of the production are manually inspected with available microscopes. Those microscopes are equipped with motorized stages and different magnifications to analyze important areas of the sample and to take care that no systematic errors occur over production time. Manual inspection is a time consuming task and puts a lot of pressure on the human inspector as documentation of this manual step is hard to accomplish and only a small amount of samples can be inspected.

We present a solution to this demands by extending a well working industrial camera setup with an automatized microscopy approach while keeping most of the established process in place but adding this microscope as another inspection unit in a modular way by reusing the given infrastructure. This installation was constructed and long-term runtime tested at Fraunhofer IOSB MicroLab.

2 Hardware and Software Setup

Our example setup consists of three core components as seen in Fig. 1a. We are instrumenting an industrial robot of Stäubli company. This TX60-L features six axis and is equipped with a pneumatic parallel gripper on the 6th axis. The robot cares for handling the specimen between industrial camera, specimen cache and microscope.

As microscopic inspection device we use a macroscope of Leica company. The Z16 APO A is equipped with a movable z-axis we are using for auto focusing. This macroscope provides continuous motorized magnification between 7.1× and 115× which is needed for detailed inspection of certain pre-selected areas. Pictures and video stream are captured by a high resolution camera with a full-frame resolution of 2048 × 1536 pixels. Because of the limited field of view when inspecting with large magnification a motorized stage of Märzhäuser, the SCAN^{PLUS} with a repeat accuracy of <1 μm was added to this inspection unit. The macroscope is instrumented by a C++ Wrapper of the low level Leica software development kit

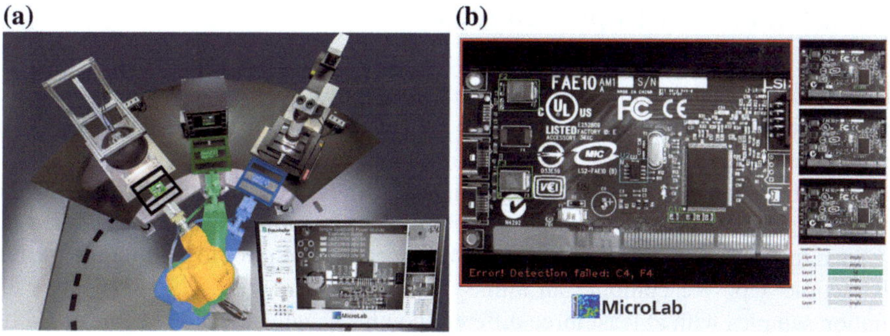

Fig. 1 Illustration of the example setup with microscope, robot, industrial camera, caching magazine (**a**) and separate multiview display (**b**)

(SDK). For the shown realization we used the light control, the z-axis, zoom control and the motorized stage.

The third component is a standard industrial gray-scale camera Manta G-145B of ALLIED Vision Technologies with a resolution of 1388 × 1038 pixels. We equipped this camera setup with a controllable 8-segment ring-light by Cool Control. To reduce external influences, like electric lighting, this unit was covered with a shield; see left side of Fig. 1a. In order to compensate for time consuming inspection at the microscope inspection unit, we added a magazine for eight specimen-samples to cache the overall process and keep throughput on a high level. The software implementation is based on a loosely coupled design. We have chosen C++ for the low-level algorithms and QT4 for realizing the message driven graphical user interface (GUI). The robot is communicating via network by maintaining an internal state-machine. The communication is done by sending messages to the robot which answers its current state together with additional information in its response-message. The robot maintains two separate processes. One is used for doing network dispatching, e.g. sending and receiving of messages and the other is controlling the internal sequence of operation by maintaining the mentioned state-machine. This approach is quite flexible as arbitrary motion sequences can be performed while the robot is caring for the programmed safety constraints byself.

The overall process is started by transporting the first specimen from the magazine onto the camera inspection unit. The camera acquires an overview image, which is used to do a completeness check. The region of interests (ROI), e.g. chips, capacitors, special circuit paths are sent to the microscope for later detailed inspection. The specimen is then transferred to the motorized stage of the microscope, where the same ROI is examined in detail. The progress of the QA cycle and the current detection results are visualized in a developed multiview video display (Fig. 1b).

3 Algorithm and Implementation

The quality measurement we are doing in this paper is based on a correlation method that matches extracted patterns of known good areas with the acquired image. Those patterns are manually extracted from either the camera's overview image or the stitched together microscope image. In order to find the matching ROI on the microscope image which corresponds to the overview image the industrial camera (*C*) and microscope's positioning stage (*M*) are calibrated prior to starting the pattern extraction step. We compute an affine transformation $\varphi : \mathscr{C} \rightarrow \mathscr{M}$ by using calibration samples with at least three different points of interest that can be resolved by the camera. As the microscope's field of view is, at decent magnification, quite small, the positioning stage needs to move this sample that all points of interest can be inspected by the microscope. With three corresponding points available the affine transformation can be computed as shown in (2). The image coordinates of the acquired camera image and the axis values of the positioning stage are donated as tuples (c_{xi}, c_{yi}), respectively (m_{li}, m_{ri}). Those tuples are written into the matrices as shown in (1). φ can be determined by computing the inverse matrix of $C \in \mathscr{C}$ and multiplying it to the right side of $M \in \mathscr{M}$. Additional corresponding points can be used to compute a more robust estimation by doing e.g. linear regression.

$$C = \begin{bmatrix} c_{x1} & c_{x2} & c_{x3} \\ c_{y1} & c_{y2} & c_{y3} \\ 1 & 1 & 1 \end{bmatrix} \quad M = \begin{bmatrix} m_{l1} & m_{l2} & m_{l3} \\ m_{r1} & m_{r2} & m_{r3} \end{bmatrix} \tag{1}$$

$$\begin{aligned} \varphi \cdot C &= M \\ \varphi &= M \cdot C^{-1} \end{aligned} \tag{2}$$

We use the normalized cross correlation coefficient $\gamma(x, y)$ as shown in (3) which was presented by Gonzalez et al. in [3]. This cross correlation coefficient is computed by moving a template $w(s, t)$ over an image $f(x, y)$. At every position (3) is evaluated. The position of the maximum value $\gamma_{\max}$ corresponds to the best matching location $p_{\gamma_{\max}}$, a tuple concerning (x, y) as shown in (4). $\gamma_{\max}$ itself is a quality criterion for the matching result. $\bar{w}$ and $\bar{f}$ denote the average value of these two functions in the given region. As one can see $\bar{w}$ is not dependent on an argument (x, y), it is therefore only calculated once.

$$\gamma(x, y) = \frac{\sum_s \sum_t [w(s,t) - \bar{w}] \sum_s \sum_t [f(x+s, y+t) - \bar{f}(x+s, y+t)]}{\sqrt{\sum_s \sum_t [w(s,t) - \bar{w}]^2 \sum_s \sum_t [f(x+s, y+t) - \bar{f}(x+s, y+t)]^2}} \tag{3}$$

$$\gamma_{\max} = \max_{x,y} \gamma(x, y) \tag{4}$$

$$p_{\gamma_{\max}} = \arg\max_{x,y} \gamma(x,y) \tag{5}$$

Within this paper we define that a positive match exists when $\gamma_{max} \geq 0.75$, which is based on our experimental results with multiple known good specimen. Illumination was maintained constant during the process and external influences were minimal, which kept this threshold on a fixed level. Figure 2 shows the result of this pattern matching on the camera's overview image. One can see that the patterns *U*1, *U*2, *C*10 are detected and e.g. the capacitor *C*4 is not. The red dashed rectangle was manually added for better comprehension. In order to check the soldering of the chips more in detail the analysis is repeated with microscopic level of magnification. A complete image with large magnification cannot be acquired with only one shot. Therefore the area needs to be assembled together by combining multiple single images. This process is called image stitching. For this algorithm we used an overlap of approximately 10 % concerning image height and image width for the acquired images. The images are stitched line by line first. Afterwards each line of images is cropped that way, that only valid image pixels are present throughout that stitched area to compensate offsets between the images that arise when more than one axis of the positioning stage is instrumented at the same time. On the next step the line of images are stitched vertically, which is done by assembling each two of the resulting lines until all temporary results have been put together.

Synthetically enhanced images, computed out of 10 input images each, according to [2] were used as input for the stitching algorithm. The images were acquired equidistantly every 0.5 mm concerning the z-axis position to compensate the small depth of view when using large magnification. Figure 3 presents the cross correlation as intermediate step and the assembled result image.

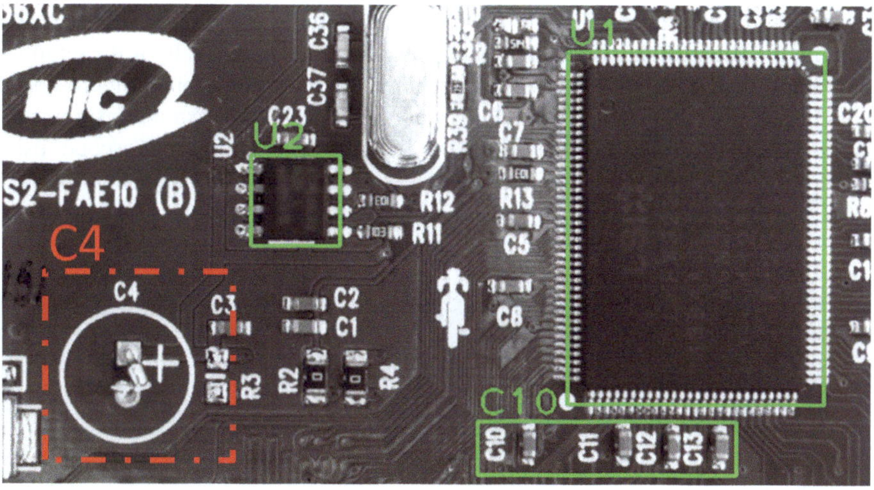

Fig. 2 The cross correlation is used to do a completeness check on the camera's gray-scale overview image. Not detected parts directly result in a failure of the QA process

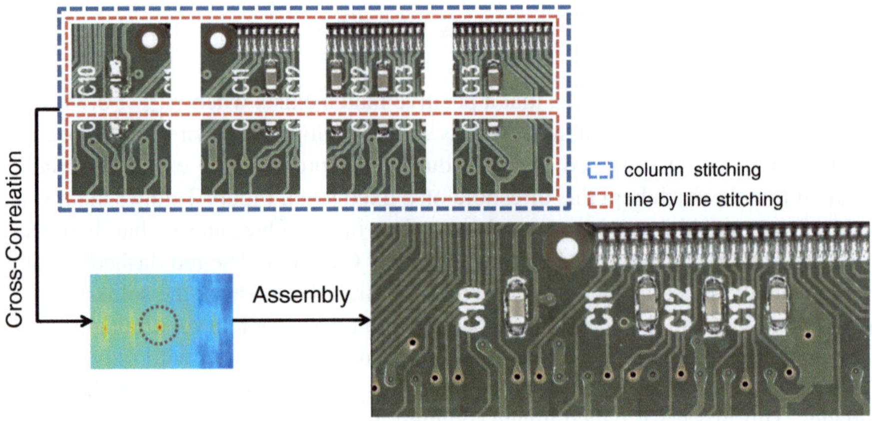

Fig. 3 Images are stitched together to generate a continuous overview. In this example 4 × 2 input images were assembled

4 Results

While the completeness check can detect missing parts on the examined specimen on the one hand, the detected availability of the smaller chips on the other hand does not proof that the established quality assurance process was successful in rejecting defective boards. In Fig. 4, we selected two examples where the comparison between the chips labeled *U*2, *C*10 on the overview image and the detailed microscopy analysis reveals a false positive on the side of the camera inspection as the chip is in case of *U*2 damaged and in case of *C*10 only partly existing.

Therefore the camera inspection system can only be used as an indicator concerning the detailed classification of small soldered chips, especially if the region of interest is represented with too few pixels. As a short term solution a camera system with higher resolution and same field of view could be used to extend this transition zone. Though this does not scale forever, a combination with a microscope, using the complement capabilities of those two systems, namely the large field of view

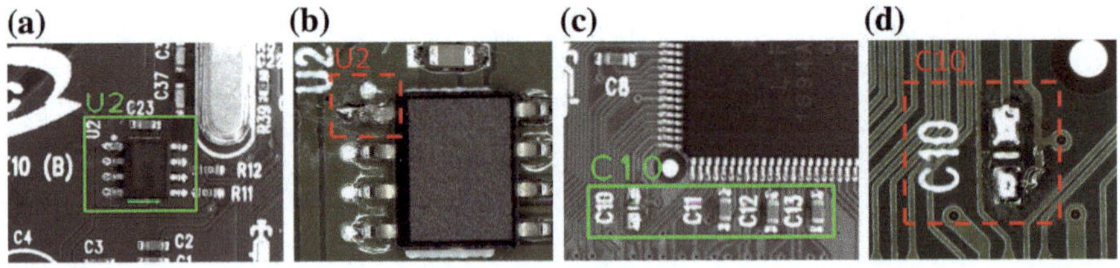

Fig. 4 Camera inspection has accepted the available chips in (**a**) and (**c**), but the microscopy inspection revealed that both parts are defective, as shown in (**b**) and in the detail of *C*10 in (**d**)

of the camera system and the large magnification of the microscope and combining them for detailed analysis is an attractive approach, that scales towards the limits of optical resolving powers.

5 Summary and Outlook

We have successfully shown a prototype using widespread industrial components in combination with a robust software implementation to extend an existing QA process. Because of the modular approach this already existing QA infrastructure can be kept as is but is enhanced with microscopic quality inspection at low costs. In case of issues with the new inline system this module can be deactivated and the old process is back in place by omitting the integrated microscope detection unit. This rollback functionality is especially important when changing an established process which has been successfully used for e.g. completeness checks for a long time. With that integration we have also demonstrated that manually operated microscopes, that are used in an existing QA process for spot-tests anyways, can be easily reused in that enhanced automated process.

We did not yet implement an easy way of teaching the patterns needed for doing the completeness check with the industrial camera or the detailed matching of the computed microscope images. This can be solved by implementing a graphical tool for selecting the to be inspected regions on the overview image when inspecting a known good part. The region, the stitched microscope images represent, is directly related to the selected region on the overview image, because of the underlying calibration between industrial camera and microscope. Therefore the pattern used for the detailed matching can be acquired without any user interaction and can be automatically extracted from the processed microscope images. The calibration, which was manually done within that paper, can also be automatized when inspecting a specimen with adequate calibration patterns by adding an extra step into this process and constantly matching the acquired image data while moving the positioning stage. The resulting correspondence points can be used to compute the affine transformation as shown in Sect. 3.

In the future we want to transfer this installation setup from the Fraunhofer IOSB MicroLab into a productive environment to measure its performance and durability under real industrial conditions.

References

1. J. Beyerer, F. Puente León, C. Frese, *Automatische Sichtprüung: Grundlagen, Methoden und Praxis der Bildgewinnung und Bildauswertung* (Springer Vieweg, Stroud, 2012)
2. P. Frühberger, E. Klaus, J. Beyerer, in *Microscopic Analysis Using Gaze-Based Interaction*, ed. by E.K. Polychroniadis, A.Y. Oral, M. Ozer. International Multidisciplinary Microscopy Congress, vol 154 (Springer International Publishing, London, 2014), pp. 195–200
3. R.C. Gonzalez, R.E. Woods, Digital Image Processing. Pearson Education (2008)

Effect of Annealing Temperature on the Structural and Magnetic Properties of Terbium Iron Garnet Thin Films Prepared by Sol-Gel Method

Ftema W. Aldbea and N.B. Ibrahim

Abstract The terbium iron garnet (TbIG) nanoparticle thin films have been prepared by a sol-gel method for potential use as a magnetic sensor. The films were deposited onto clean quartz substrate using a spin coating technique, followed by annealing at different temperature of 700, 800, 900 °C in air for 2 h. As prepare film has been also executed. Microstructural and magnetic properties of the films were measured using an X-ray diffraction (XRD), a Field Emission Scanning Electron Microscope and a Vibrating Sample Magnetometer. The XRD results revealed that the pure phase of garnet structure has been formed at 900 °C. The lattice parameter decreased initially from 11.6 to 11.4 Å when annealing temperature increased from 700 to 800 °C, and then it is increased to 12.36 Å when annealing temperature increased to 900 °C. The grain sizes increased from 12 to 18 nm with increasing of annealing temperature from 700 to 900 °C, while the thickness of the films decreased to the smallest value of 312 nm at 900 °C. The saturation magnetization increased with increasing of annealing temperature due to increasing of grain sizes. The coercivity of films exhibited a highest value of 102 kA/m at 900 °C.

1 Introduction

Terbium iron garnets ($Tb_3Fe_5O_{12}$) are an indispensible class in garnet ferrite group due to their potential in applications such as, storage, communications and magnetic sensors [1, 2]. The crystal structure of TbIG is a cubic and belongs to Ia 3d–(O_h^{10}) group; the Tb^{3+} ions are occupied the dodecahedral sites {24c}(222) while the two Fe^{3+} ions are distributed in the octahedral [16a](3) and tetrahedral (24d)(4) sites, hence the magnetic formula of TbIG can be written as $\{Tb_3^{+3}\}[Fe_3^{+2}](Fe_3^{+3})O_{12}$ [3].

F.W. Aldbea (✉) · N.B. Ibrahim
Faculty of Science and Technology, Universiti Kebangsaan Malaysia, Bangi, Malaysia
e-mail: ftemaaldbea@yahoo.com

F.W. Aldbea
Faculty of Science, Sebha University, Sebha, Libya

E.K. Polychroniadis et al. (eds.), *2nd International Multidisciplinary Microscopy and Microanalysis Congress*, Springer Proceedings in Physics 164,
DOI 10.1007/978-3-319-16919-4_9

Terbium iron garnets have attracted more attention because it exhibited a ferrimagnetic ordering of the Tb and iron moments along the [111] direction at room temperature, whereas noncollinear magnetic structures are observed at low temperatures [4–6]. Few Authors have reported that terbium iron garnet thin films prepared by pulsed laser deposition technique showed anomalously large coercivity and in–plane easy axis of magnetization. In addition, the M-H loop measured at 20 K, exhibited about an order of magnitude enhancement in the coercivity value (H_c) than the single crystal. In addition, easy axis of the magnetization lies within film's plane [2, 7]. No report so far for TbIG thin films prepared by simple steps for sol gel technique. This work reports the effect of annealing temperature on the structural and magnetic properties of TbIG. The sol-gel method is preferable in this work to produce thin films because it offers considerable advantages, like better homogeneity, capability to produce nanoparticles samples, simple preparation method and low preparation temperature [8, 9]. It is a wet chemical method when the solid precursor is dissolved into a solvent.

2 Method

2.1 Samples Preparation

The terbium solution was prepared by dissolving terbium (III) acetatehydrate ($Tb(COOCH_3)_3.H_2O$, Alfa Aesan, 99.9 % purity) in 2 mL of acetic acid, 1 mL of distilled water and 5 μL of nitric acid and iron (III) nitrate nonahydrate ($Fe(NO_3)_3 \cdot 9H_2O$, 98.0 % purity) was dissolved separately in 2 mL of 2-metha-oxythanol. The Tb–Fe solution refluxed at (80 ± 1) °C for 3 h. The precursor solution was maintained at pH 3 by adjusting the amount of diethylamine around 0.2 mL [10, 11]. The resulting solution was refluxed again for 3 h at (80 ± 1) °C followed by stirring for 48 h to obtain the homogenous gel. To prepare the film, 25 μL of the gel was dropped onto a clean quartz substrate (1.45 mm thickness). The substrate was spin coated first at 500 rpm for 15 s and then at 3500 rpm for 30 s. The film was heated at 90 °C to remove the residual solvent and at 350 °C for 15 min to burn-off the organic compounds. Then, the annealing process was carried out at 700, 800 and 900 °C in air for 2 h.

2.2 Characterization Techniques

The films nanostructures were characterized using an X-ray diffractometry (XRD, D8 Advance Bruker) with CuKα radiation (λ = 0.15406 nm) in a 2θ range of 20°–60°. The surface morphology of the films was examined using a field emission scanning microscope (FE-SEM, ZEISS Supra 55VP) operating in vacuum at 15 kV using backscatter detection. The films were coated with gold to prevent the accumulation of electrostatic charges at the surface to obtain a good image. Carbon tape was used to stick the film onto the sample holder in FE-SEM. In-plane magnetic measurements at room temperature (25 °C) were carried out using a vibrating sample magnetometer (VSM, Lake Shore 7404) with a maximum magnetic field value of 16,000 kA/m. The films were cut into small pieces of 4 × 4 mm^2. The magnetization loops detected by VSM represented the magnetization of film and the substrate. No net ferrimagnetic sign was observed from the raw data, because of the diamagnetic effect from the quartz substrate. Hence, the diamagnetic contribution has been corrected for all of the hysteresis loops of the films. This correction eliminated the contribution of the substrate.

3 Results

3.1 Structural Analysis

Figure 1 shows the XRD patterns of $Tb_3Fe_5O_{12}$ films as prepared and annealed at 700, 800 and 900 °C. The results were compared with the *Joint Committee on Powder Diffraction Standard* (JCPDS-ICDD; number 01-071-0697) for pure TbIG. The films before annealing demonstrate an amorphous structural behavior. At 700 and 800 °C, the films show some of garnet peaks mixed with the minute quantity of Fe_2O_3 and Fe_3O_4, because of those annealing temperature were inadequate to complete the crystallization [12]. At 900 °C, the film shows a pure phase of garnet structure resulted a shifting of diffraction angle to the lower 2θ angle.

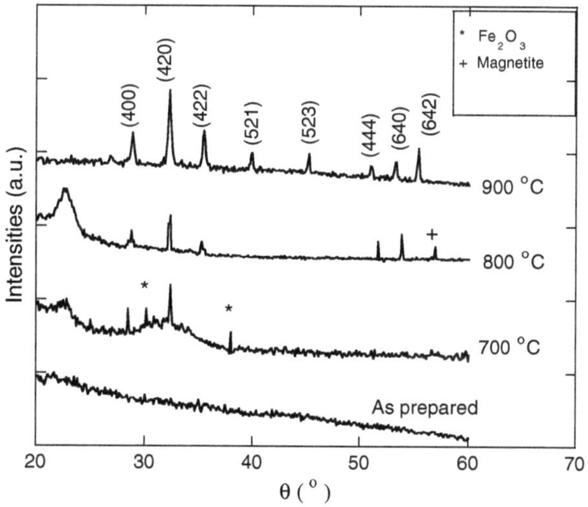

Fig. 1 The XRD patterns of TbIG films

Fig. 2 Variations of lattice parameter with annealing temperature

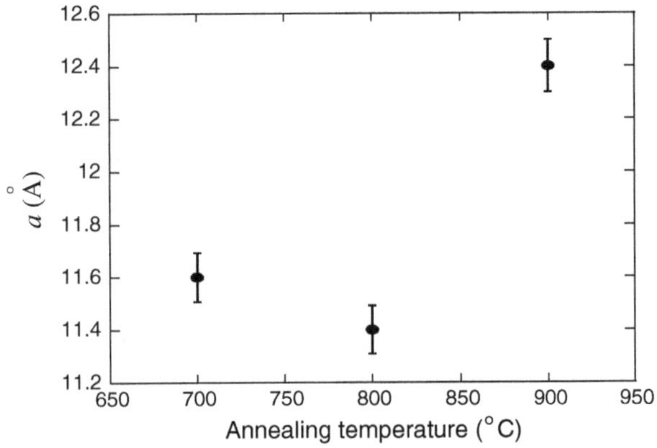

The lattice parameter *a* of the films at different annealing temperatures values of 700, 800 and 900 °C was calculated from the XRD patterns by using *Bragg's* law. *Bradley and Jay's* extrapolation against cos2θ [13] was used to calculate the lattice parameter of phase present at different content. The lattice parameters values of films at annealing temperatures between 700 and 900 °C are illustrated in Fig. 2. It can be seen that the lattice parameter of the films is responded to increment of annealing temperature. The films show a highest lattice parameter value of 12.36 Å at 900 °C which is close to 12.4 Å for TbIG bulk [3], but it is lower than 12.59 Å for the film which was prepared by pulsed laser deposition technique [7]. Increasing of the lattice parameter initiated the shifting of the diffraction peaks to the lower 2θ angle [14].

3.2 Morphological Characteristic

The FE-SEM images the TbIG films are shown in Fig. 3. The films exhibited agglomerate grains form with voids. The grains size of films has observed to be increased as annealing temperature is increased. The bigger grain size value for film was recorded to be 18 nm at 900 °C due to increase of surface energy [15]. However, increasing the grains size of films with annealing temperature could affect the magnetic properties. The thickness of films was measured from the film's cross section backscattered FE-SEM micrograph. Figure 4 shows the thickness images of TbIG films. The average thickness values of TbIG films are found to be decreased as 456, 344 and 312 nm at annealing temperature of 700, 800 and 900 °C, respectively. It can be observed that the film with bigger grain size exhibited reduction in the thickness due to the agglomeration.

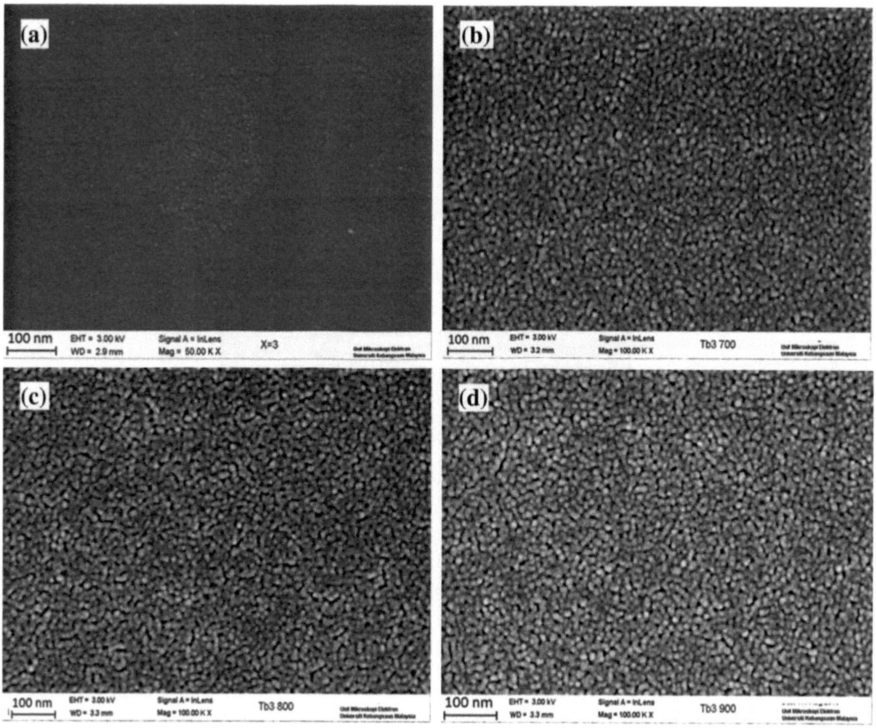

Fig. 3 The FE-SEM images of films **a** as prepared and annealing at **b** 700 °C, **c** 800 °C and **d** 900 °C

3.3 Magnetic Properties

In plan hysteresis loops of TbIG films at different annealing temperatures of 700, 800 and 900 °C are shown in Fig. 5. It can be observed that all films were saturated and the hysteresis loops shapes indicate the soft ferrimagnetic nature of films. The M_s and H_c at room temperature (25 °C) were obtained from the hysteresis loop of each film. Figure 6 shows variations of M_s and H_c of TbIG films with annealing temperature. From the Figure it can be seen that at 700 and 800 °C the films have same M_s value (3 kA/m); low M_s could be related to existence of Fe_2O_3 and Fe_3O_4 in the films. Increasing of annealing temperature up to 900 °C led to increase M_s to 13 kA/m which is smaller than 23 kA/m for TbIG powder prepared by sol-gel method [16]. Additionally, evolution of the grain size with annealing temperature causes to increase the M_s [17]. The H_c of films has observed initially decreases from 135 to 67 kA/m as annealing temperature increases from 700 to 800 °C, but it is increased markedly at 900 °C with value of 102 kA/m, the increasing of H_c can be

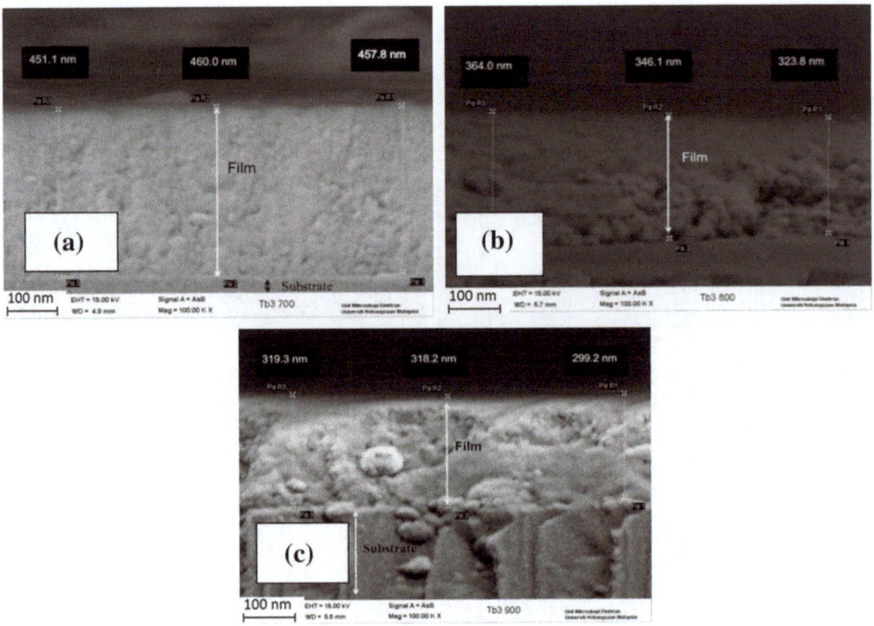

Fig. 4 The cross sectional images of films annealed at **a** 700 °C, **b** 800 °C and **c** 900 °C

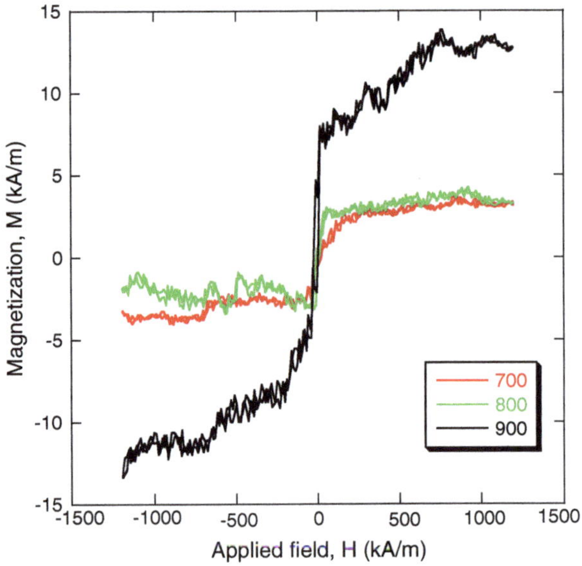

Fig. 5 The hysteresis loops of TbIG films annealed at 700, 800 and 900 °C

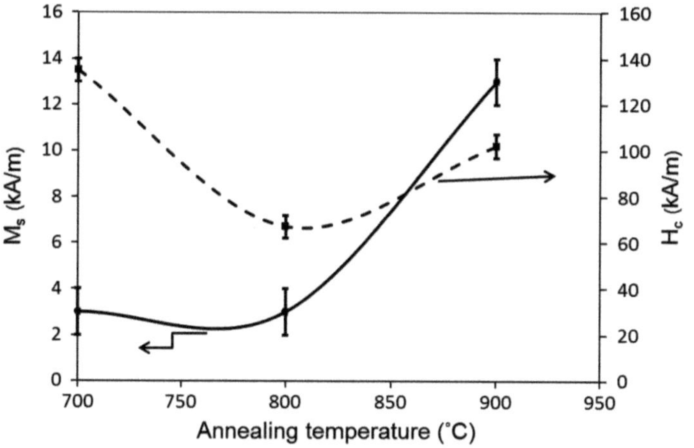

Fig. 6 Variations of M_s and H_c of films with annealing temperature

stated to the coercivity of nanoparticles that is governed by permanent domain rotation [18]. Although the H_c of films shows a highest value at 900 °C, but still has a soft magnetic feature because of it is less than 1000 kA/m.

4 Conclusions

Effect of annealing temperature on the structural and magnetic properties of TbIG sol-gel films has been studied. The results indicated that best annealing temperature to crystalline the films is 900 °C. The M_s and H_c of films are also responded to increment of annealing temperature. The films have potential in magnetic sensor and microwave applications due to low H_c. However, the films showed an interesting result at different annealing temperature values which is need more study in further work.

Acknowledgments The authors would like to thank Malaysian Ministry of Science and Technology for supporting this work under Grant No. 03-01-02-SF0538.

References

1. R.J. Joseyphus, A. Arayanasamy, A.K. Nigam, R. Krishnan, Effect of mechanical milling on the magnetic properties of garnets. Magn. Magn. Mater **296**, 57–64 (2006). doi:10.1016/j.jmmm.2005.04.018
2. N. Kumar, R. Bhargava, S. Kumar, A.K. Chawla, R. Chandra, M. Bohra, In-plane magnetization with high coercivity in terbium iron garnet thin films deposited on Pt/Si

substrate by PLD. Magn. Magn. Mater **322**, 1727–1730 (2010). doi:10.1016/j.jmmm.2009.12.015
3. M. Lahoubi, M. Guillot, A. Marchand, F. Tcheou, E. Roudault, M. Lahoubi, M. Guillot, A. Marchand, F. Tcheou, E. Roudault, Double umbrella structure in terbium iron garnet. IEEE Trans. Magn. **20**(5), 1518–1520 (1984). doi:10.1109/TMAG.1984.1063259
4. S. Thongmee, P. Winotai, I.M. Tang, Local field fluctuations in the substituted aluminum iron garnets $Y_3Fe_{5-x}Al_xO_{12}$. Solid State Commun. **109**, 471–476 (1999). doi:S0038-1098(98)00583-3
5. R. Hock, H. Fuess, T. Vogt, M. Bonnet, Crystallographic distortion and magnetic structure of terbium iron garnet at low temperatures. Solid State Chem. **84**(1), 39–51 (1990). doi:10.1016/0022-4596(90)90182-W
6. S. Geller, J. Remeika, R. Sherwood, H. Williams, G. Espinosa, Magnetic study of the heavier rare-earth iron garnets. Phys. Rev. **137**, 1034–1038 (1965). 103. http://dx.doi.org/10.1103/PhysRev.137.A1034
7. N. Kumar, N.G. Kim, Y.A. Park, N. Hur, J.H. Jung, K.J. Han, K.J. Yee, Epitaxial growth of terbium iron garnet thin films with out-of-plane axis of magnetization. Thin Solid Films **516** (21), 7753–7757 (2008). doi:10.1016/j.tsf.2008.05.032
8. X. Haitao, H. Yang, W. Xu, L. Yu, Magnetic properties of Bi-doped $Y_3Fe_5O_{12}$. Current App. Phys. **8**, 1–5 (2008). doi:10.1016/j.cap.2007.04.002
9. S. Hossini, H.R. Madaah, Z.A. Nemati, Synthesis of nanocrystalline yttrium iron garnet by sol-gel combustion process: the influence of pH of precursor solution. Mater. Sci. Eng. B. **129**, 211–215 (2006). doi:10.1016/j.mseb.2006.01.014
10. F.W. Aldbea, N.I. Ahmad, N.B. Ibrahim, M. Yahya, Effect of increasing pH value on the structural, optical and magnetic properties of yttrium iron garnet films prepared by a sol-gel method. Sol-Gel Sci. Technol. **71**(1), 31–33 (2014). doi:10.1007/s10971-014-3326-4
11. F.W. Aldbea, N.B. Ibrahim, M.H. Abdullah, R.E. Shaiboub, Structural and magnetic properties of $Tb_xY_{3-x}Fe_5O_{12}$ (x = 0 to 0.8) thin film prepared via sol-gel method. Sol-Gel Sci. Technol. **62**, 483–489 (2012). doi:10.1007/s10971-012-2752-4
12. A.C. Rastogi, V.N. Moorthy, Magnetic properties of yttrium iron garnet thin films changed through surface modification by CoO overlayer growth for magneto-optic recording applications. Magn. Magne. Mater **214**, 228–239 (2002). doi:10.1016/S0304-8853(01)01360-9
13. V.K. Pecharsky, P.Y. Zavalij, *Fundamentals of Powder Diffraction and Structural Characterization of Metals* (Springer, New York, 2009. doi:10.1007/978-0-387-09579-0
14. N.B. Ibrahim, C. Edwards, S.B. Palmer, Pulsed laser ablation deposition of yttrium iron garnet and cerium-sbstituted Y films. Magn. Magn. Mater **220**, 183–194 (2000). doi:10.1016/S0304-8853(00)00331-0
15. F.W. Aldbea, N.B. Ibrahim, M.H. Abdullah, R.E. Shaiboub, *Effect of Annealing Temperature on the Structural and Magnetic Properties of* $TbxY_{3-x}Fe_5O_{12}$(x = 0.0, 1.0, 2.0) *Thin Films Prepared by Sol-Gel Process* (RCSSST, Malaysia, Trans Tech Publications, Switzerland, 2002). doI:10.4028/www.scientific.net/AMR.501.236
16. H. Young Jun, K. Jun Sig, S. In-Bo, K. Chul Sung, Spin rotation at compensation point studies of $Tb_3Fe_5O_{12}$ by Mössbauer spectroscopy. Magn. IEEE Trans. Magn. **40**(4), 2808–2810 (2004). doi:10.1109/TMAG.2004.832105
17. T.-C. Mao, J.-C. Chen, Influence of the addition of CeO_2 on the microstructure and the magnetic properties of yttrium iron garnet ceramic. Magn. Magn. Mater. **302**, 74–81 (2006). doi:10.1016/j.jmmm.2005.08.018
18. X. Haitao, H. Yang, L. Lu, Effect of erbium oxide on synthesis and magnetic properties of yttrium iron garnet nanoparticles in organic medium. Mater. Sci. Mater. Electron. **19**, 509–513 (2008). doi:10.1007/s10854-007-9372-8

Fibrous Growth of Chloride Minerals on Diatomite Saturated with a Brine

Tomasz Tobola, Marek Rembiś, Beata Figarska-Warchoł and Grażyna Stańczak

Abstract One of the fundamental characteristics of diatomites is their structural porosity as it controls specific properties of these rocks and their possible industrial applications. The authors examined the interaction between natural diatomites (not calcinated) from Polish Carpathian Flysch Belt and a brine composed of the mixture of NaCl, KCl and $MgCl_2*H_2O$ with the concentration 100 g/L each for 5 days. Blocks of diatomites immersed in the brine showed an uprising boundary of a moisturised rock after a 2–4 h. Then, after about 24 h salt efflorescences began to grow on their surfaces. SEM-EDS analyses have revealed two types of fibres on the block surfaces. One group is formed by long and thin, often strongly bent fibres. In the cross-section they are rounded or flatten (ribbons) or have more complex shapes. Their chemical composition indicates sylvine but with a considerable amount of sodium. The crystals of the second type are shorter and thicker, and their chemical composition indicates halite but with high amounts of potassium. The magnesium minerals was not stated. Inside the lower parts of diatomite blocks, pores are almost completely filled with halite, but in the upper parts of the blocks sylvine is the precipitate found in the pores. The habit of the efflorescent salts seem to be influenced by the size and shape of the diatomite porous structure. The spatial distribution of both salt types and their chemical compositions seem to be associated with selective absorption of ions by silica that forms the diatomite framework.

1 Introduction

Diatomite is a siliceous rock composed of skeletal remains of single-cell diatoms cemented by opal that may be recrystallized into chalcedony and microquartz. It has high total porosity and low volume density, a porous and permeable structure,

T. Toboła (✉) · M. Rembiś · B. Figarska-Warchoł · G. Stańczak
Faculty of Geology, Geophysics and Environmental Protection, AGH University of Science and Technology, al. Mickiewicza 30, 30-059 Kraków, Polska
e-mail: tob@geol.agh.edu.pl

E.K. Polychroniadis et al. (eds.), *2nd International Multidisciplinary Microscopy and Microanalysis Congress*, Springer Proceedings in Physics 164,
DOI 10.1007/978-3-319-16919-4_10

chemical durability, large pore area, high adsorption capacity and good adsorptive properties. These features are useful in many industrial applications of this raw material.

The purpose of the study was to establish the types of salt forms developed as a result of interaction between diatomites and highly saturated brines. We used a multicomponent brine to determine the selective absorption of Na^+, K^+, Mg^{2+}, Cl^- ions by diatomites.

2 Materials and Methods

The material was taken from the Jawornik Ruski deposit of Lower Miocene diatomite (SE Poland, the Outer Carpathians, Leszczawka syncline). Several types of diatomites have been distinguished in the deposit: blocky, platy, prismatic, nodular, shale-clayey and silicified ones [1]. The investigations were carried out on two rock types: light creamy, soft, with water absorption [2] 27.6–36.2 %, and grey-brown, hard, with water absorption 13.0–26.9 %. Both rock types have blocky or platy parting.

Irregular, several centimetres large fragments of two selected types of diatomites were put into a brine which initially reached about 10 % of their height. The brine was composed of 100 g NaCl, 100 g KCl and about 100 g $MgCl_2{*}H_2O$ dissolved in 1 L of distilled water. Such prepared samples were stored in the ambient air at the temperature 20 °C for 5 days. After that time the specimens were dried under the same conditions to a constant weight. Then small pieces of diatomites were taken from lower parts of the samples (i.e. the parts that had been near the brine level) and from the parts previously situated ca. 5 cm above the brine level. These rock pieces were used in SEM-EDS analyses in order to determine the forms and mineral composition of the efflorescent salts as well as the manner and extent the salt filled the pore space. The samples were prepared with carbon sputtering. The SEM-EDS analyses were performed using a FEI Quanta 200 FEG scanning microscope equipped with an EDS detector. The microscope worked in high vacuum at a current of 20 kV. The mineral composition of efflorescent salts was analysed with the powder X-ray diffraction (XRD) method using a Philips X'Pert APD diffractometer with a PW 3020 goniometer, a Cu lamp and a graphite monochromator. The X-ray analyses were carried out within the range 2–72°2Θ.

3 Results

On the surface of the diatomites soaking in the brine, the moving up boundary of moisture was clearly visible after a few hours and after 24 h first manifestations of salt efflorescences appeared. The rate of soaking as well as the intensity of salt efflorescence were different on two types of diatomites. After 5 days the most

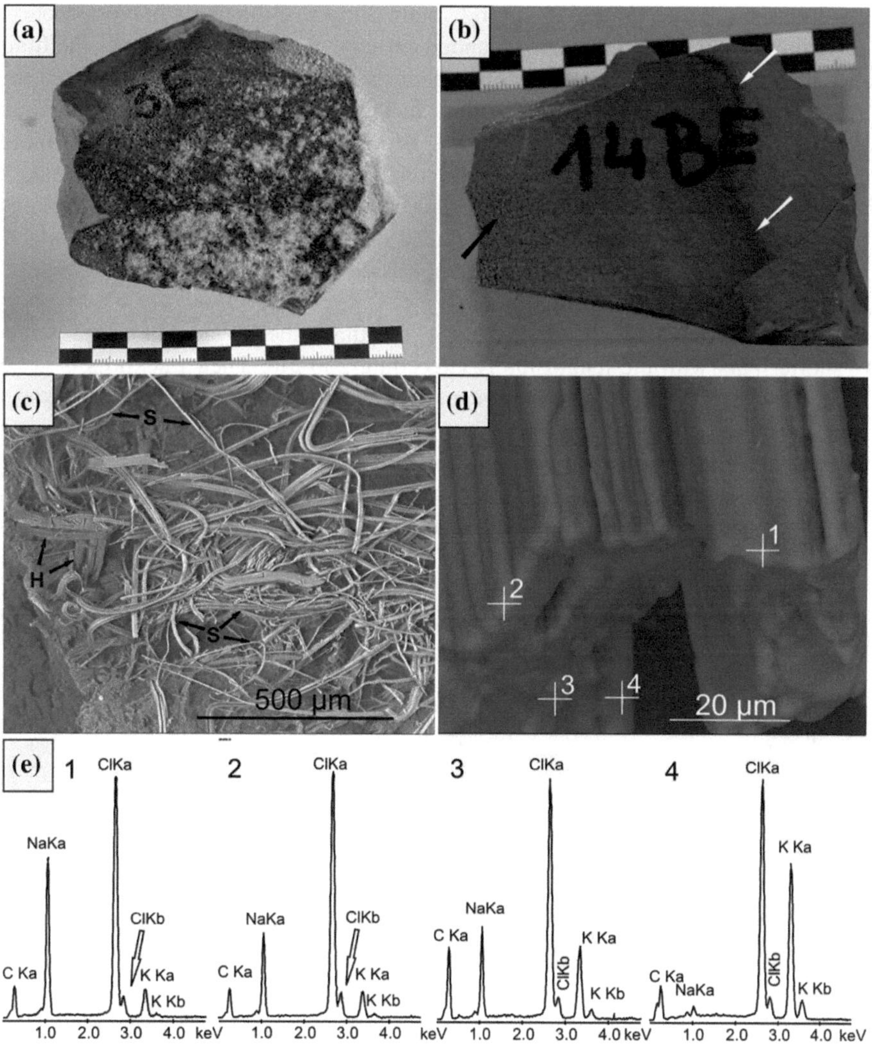

Fig. 1 **a** Efflorescence of salts on the surface of the light creamy diatomite. **b** Moisture area (*white arrows*) with poor salt efflorescence (*black arrow*) on the surface of the *grey-brown* diatomite. **c** Fibres of "sylvine" (*S*) and "halite" (*H*) on the surface of diatomite in an SEM-EDS image. **d** Sub-fibres forming elongated fibres. *Crosses* mark measurements points (EDS). **e** Set of chemical analyses (EDS) in points 1–4 of the photo **(d)**

densely packed fibres of salt were noticed on the light creamy diatomites (Fig. 1a). On the contrary, the grey-brown diatomites had a rare and short fibres and the moisture line was very well visible (Fig. 1b).

On the surface of the light creamy diatomites two types of fibres have been distinguished (Fig. 1c). The first of them are long (up to a few millimetres) and thin

(20–30 μm) and often curved or strongly bent. Shapes of the fibres in cross-sections vary from irregular polygons with rounded corners to circular. Sometimes the fibres are flatten and form a twisted ribbon. Chemical microanalysis (EDS) showed that this type consist of K, Na and Cl ions. In the majority of the fibres potassium prevails over sodium indicating "sylvine" as a main phase and "halite" as a minor phase. Many of these fibres show more complex structure at higher SEM magnifications (Fig. 1d). They consist of parallel, very thin (a few μm), densely packed, rounded in cross-section sub-fibres. Detailed analyses of single sub-fibres indicate that the K/Na ratio is highly variable: potassium usually prevails but sometimes sodium dominate (Fig. 1d, e).

The second form of efflorescences on the diatomite surfaces are represented by shorter bars with the length rarely exceeding 250 μm. Their thickness is larger (up to 100 μm) and shape more regular; the bar corners are usually rounded (Fig. 2a). The chemical microanalyses (EDS) indicate Na and Cl being main components, sometimes accompanied by K (Fig. 2b).

The porous space of the diatomite samples is almost completely filled with the "halite" phase at the lower part (near the former brine level). The "sylvine"-like

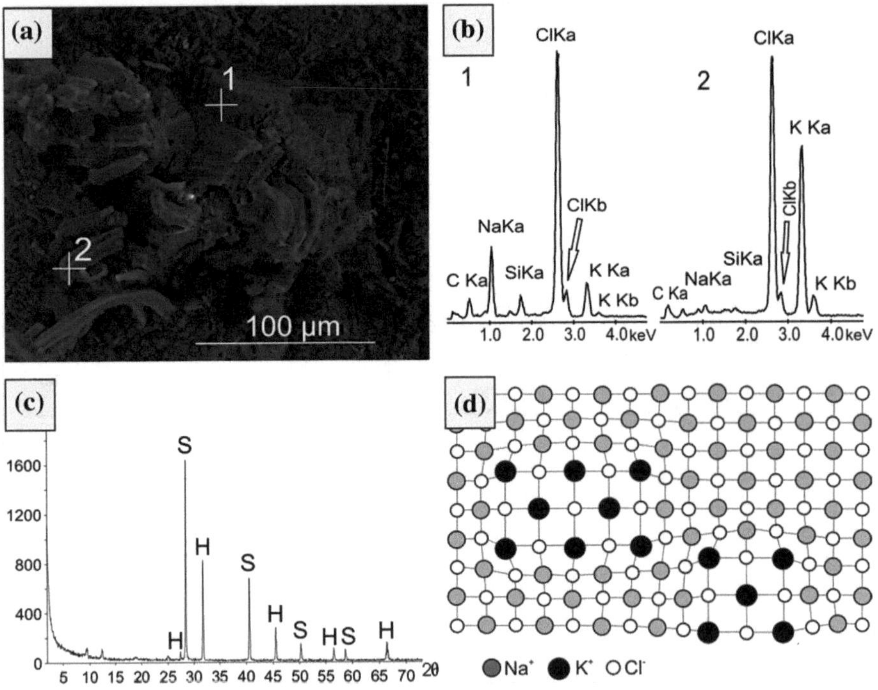

Fig. 2 **a** Short, "halite" fibres with rounded corners (point *1*) and "sylvine" fibres (point *2*). **b** Set of chemical analysis in points *1* and *2*. **c** X-ray pattern of fibres from the surface of diatomite (*H* halite, *S* sylvine). **d** A sketch of probable structure of "halite" or "sylvine" fibres

fibres appear in the upper parts of samples (about 5 cm above the former brine level). In both cases the chemical and mineral compositions are similar to those of the respective surface fibres.

4 Discussion and Conclusions

Salt efflorescences on the surface of diatomites as well as in their porous space reveal atypical features both in the respect of shapes and a chemical composition. The salt efflorescences do not have the cubic shape of natural halite or sylvine but form two types of fibres: (1) thin and long, and (2) thicker and shorter. Both of them differ in chemical and mineral compositions. According to Zehnder and Arnold [3], such forms grow during the crystallization from small volumes of solutions (phase 3–4, fibres and whiskers growth [3]). In our experiment the brine was transported through connected pores via the capillary mechanism, so the brine saturation and salt crystallization took place from thin films of the solution both on rock surfaces and on the walls of pores. The problem of precipitation of salt under such conditions seems to be more complex and not only depend on a volume of a solution but also on a type of the substrate, i.e. diatomite porosity and mineral composition. Using the four-component solution (K, Na, Mg, Cl) one could expect precipitation of halite, sylvine and carnallite. However, no magnesium minerals have been found on the surface of the diatomites and in their porous space. This may indicate that magnesium is not adsorbed by diatomite due to a selective adsorption of the rock.

The "sylvine" and "halite" fibres also have atypical compositions. Under natural conditions (i.e. in salt formations) both minerals precipitate simultaneously from saturated brines and at low (ambient) temperatures they do not form solid solutions, which is possible at higher temperatures [4–6]. At low temperatures the substitution of potassium in halite and sodium in sylvine is negligible [7–10]. On the contrary, both types of fibres occurring on the surface and in porous space of the diatomites represented mixtures of KCl and NaCl, with KCl prevailing in the long fibres and NaCl in the shorter ones. In both cases, the XRD analyses showed the presence of two separate mineral phases (halite and sylvine—Fig. 2c), which were not revealed in the SEM-EDS analyses in single sub-fibres or bigger fibres, even at high magnifications (Fig. 1d). It is probable that the fibres are composed of very small, nano-sized crystals of NaCl and KCl bond together as a result of coherent precipitation. Such a structure of the fibres results in a formation of many bond defects between halite and sylvine nanocrystals (Fig. 2d). Their presence is confirmed by relatively wide and low peaks at 9.46 and 12.39°2θ in X-ray diffraction patterns (Fig. 2c). The mechanism of such a salt crystal growth and the interaction between silicate components of diatomite and the simple ions in brine solution remains unknown. However, the study has shown that diatomite properties highly affect the habit and chemical (mineralogical) composition of salt efflorescences on the surface of diatomites and within their porous space.

Acknowledgments The study was financed by the AGH-UST statutory grant No. 11.11.140.320.

References

1. J. Kotlarczyk, M. Brożek, M. Michalski, Diatomites of the polish carpathians—occurrence, quality, modifications and applications. Gospodarka Surowcami Mineralnymi **2**(3–4), 497–523 (1986) (summary in English)
2. EN 13755:2008, *Natural Stone Test Methods. Determination of Water Absorption at Atmospheric Pressure*
3. K. Zehnder, A. Arnold, Crystal growth in salt efflorescence. J. Cryst. Growth **97**, 513–521 (1989)
4. J.B. Thompson, D.R. Waldbaum, Analysis of the two-phase region halite-sylvite in the system NaCl-KCl. Geochim. Cosmochim. Acta **33**, 671–690 (1969)
5. D. Walker, P.K. Verma, L.M.D. Cranswick, R.L. Jones, S.M. Clark, S. Buhre, Halite-sylvite thermoelasticity. Am. Mineral. **89**, 204–210 (2004)
6. D. Walker, P.K. Verma, L.M.D. Cranswick, S.M. Clark, R.L. Jones, S. Buhre, Halite-sylvite thermoconsolution. Am. Mineral. **90**, 229–239 (2005)
7. H. Borchert, R.O. Muir, Salt deposits. The origin, metamorphism and deformation of evaporites. (D. Van Nostrand Company Ltd., London, 1964), pp. 1–338
8. R. Kühn, Geochemistry of German potash deposits, in *Saline Deposits*, ed. by R. B. Mattox. (The Geological Society of America, New York, 1968), Spec. Pap. **88**, 427–504
9. O. Braitsch, *Salt Deposits. Their Origin and Composition* (Springer-Verlag, New York, 1971), pp. 1–297
10. W.T. Holser, Trace Elements and Isotopes in Evaporites, in *Marine Minerals. Short Course Notes*, vol. 6, ed. by R.G. Burns (Mineralogical Society of America, Washington, 1979), pp. 295–346

Fractal Characteristics of the Pore Network in Diatomites Using Mercury Porosimetry and Image Analysis

Grażyna Stańczak, Marek Rembiś, Beata Figarska-Warchoł and Tomasz Toboła

Abstract The complex pore space considerably affects the unique properties of diatomite and its significant potential for many industrial applications. The pore network in the diatomite from the Lower Miocene strata of the Skole nappe (the Jawornik deposit, SE Poland) has been investigated using a fractal approach. The fractal dimension of the pore-space volume was calculated using the Menger sponge as a model of a porous body and the mercury porosimetry data in a pore-throat diameter range between 10,000 and 10 nm. Based on the digital analyses of the two-dimensional images from thin sections taken under a scanning electron microscope at the backscattered electron mode at different magnifications, the authors tried to quantify the pore spaces of the diatomites using the box counting method. The results derived from the analyses of the pore-throat diameter distribution using mercury porosimetry have revealed that the pore space of the diatomite has the bifractal structure in two separated ranges of the pore-throat diameters considerably smaller than the pore-throat sizes corresponding to threshold pressures. Assuming that the fractal dimensions identified for the ranges of the smaller pore-throat diameters characterize the overall pore-throat network in the Jawornik diatomite, we can set apart the distribution of the pore-throat volume (necks) and the pore volume from the distribution of the pore-space volume (pores and necks together).

1 Introduction

The microstructural properties of the pore space in sedimentary rocks (e.g. sandstones, diatomites, limestones, etc.) affect many pore-level processes that control fluid flow, crucial for oil and gas recovery from reservoir rocks, transport of ground water and pollutant migration [1–3]. Quantification of pore network structure is a

G. Stańczak (✉) · M. Rembiś · B. Figarska-Warchoł · T. Toboła
Faculty of Geology, Geophysics and Environmental Protection, AGH University of Science and Technology, al. A. Mickiewicza 30, 30-059 Krakow, Poland
e-mail: gstanczak@geol.agh.edu.pl

E.K. Polychroniadis et al. (eds.), *2nd International Multidisciplinary Microscopy and Microanalysis Congress*, Springer Proceedings in Physics 164,
DOI 10.1007/978-3-319-16919-4_11

critical task if there is a need to relate these characteristics to pore-scale mechanisms of transport because the rock porosity has been recognized as a complex and irregular pattern that can be considered also as a chaotic structure [4–8]. The application of the fractal geometry to quantify the spatial heterogeneities of pore space over a wide range of pore-sizes in sedimentary rocks allows to extract fractal dimension, being a structural parameter of the roughness of the pattern or of the surface irregularity, describing a heterogeneous feature with a non-integer value [9, 10].

This study presents the fractal dimensions of pore network calculated from (i) mercury porosimetry data using the Menger sponge as a model of a porous body and (ii) processed images of thin sections taken under a scanning electron microscope with the box counting method. Samples of diatomite come from the Lower Miocen diatomite horizon occurring in the core part of the Leszczawka syncline and were collected from the diatomite deposit in Jawornik Ruski in the SE Poland (the Outer Carpathians) [11]. Diatomite is a pale-colored, soft, fragile, very fine-grained and light-weight siliceous sedimentary rock, formed by the accumulation and compaction of the opaline frustules of diatoms, along with additional components, i.e. siliceous sponge spicules, clay minerals and detrital grains. Diatoms, as unicellular phytoplankton (algae), have specific micro- and nanoporous structures of opaline outershells, called frustules, and frustule dimensions vary from less than 1 to more than 100 μm depending on their species [12, 13].

2 Materials and Methods

2.1 Samples and Thin-Section Images

Samples included five diatomites of different color from white and pale grey to brownish. They have a relatively narrow range of densities (2.22–2.33 g/cm^3) measured with a helium pycnometer (a Micromeritics Accupyc 1330) whereas apparent densities determined with a mercury intrusion porosimeter (a CE Instruments Pascal 140/240) include a wider range (2.07–2.27 g/cm^3). Total porosities calculated using the data from both methods vary widely from 33.66 to 42.99 % (Table 1). One

Table 1 Densities and porosities of the Jawornik diatomites

Sample	1	2	3	4	5
Colour	White	White	Pale-grey	Brownish	Pale-brownish
Bulk density[a] (g/cm^3)	1.28	1.38	1.45	1.39	1.55
Apparent density[a] (g/cm^3)	2.07	2.23	2.18	2.08	2.27
Density (AccuPyc) (g/cm^3)	2.24	2.30	2.28	2.22	2.33
Effective porosity[a] (%)	38.19	38.17	33.68	33.34	31.89
Total porosity (%)	42.99	40.07	36.54	37.42	33.66

[a]Measured with a mercury porosimeter
(AccuPyc) measured with a helium pycnometer

polished thin section impregnated with blue-dyed epoxy was made per sample. All thin sections were examined by scanning electron microscopy, using a FEI Quanta 200F (a field emission scanning electron microscope) under high vacuum conditions in the backscattered electron mode (BSE). The field emission SEM was operated at an accelerating voltage of 20 kV and a working distance of 9.7–10.1 mm.

2.2 Image Analysis and Fractal Dimension

Images of the thin sections acquired with the SEM at the BSE mode were digitized as 1024 × 943- and 2048 × 1886 8-bit grey-scale micrographs depending on magnification: approximately 1000–1200× (the length of a scale bar 100 μm) and 6000× (the length of a scale bar 20 μm), respectively. For every sample of diatomite two representative micrographs, at the lower and higher magnification, were chosen to image analysis.

The SEM images were processed to extract the pore space and its properties using the ImageJ software [14]. After calibration based on the scale bar, each image was cropped to the area occupied by the studied sample. Thus, the 100 μm-scale images (1024 × 882) cover an area of approximately 226–282 × 195–243 μm (44,000–69,000 μm^2) depending on the pixel size (0.221–0.276 μm), whereas the 20 μm-scale images have almost a constant pixel size of 0.021–0.024 μm and include an area of nearly 43–50 × 37–43 μm (1,577–2,133 μm^2). The binarization of the images into a pore phase and a solid phase was a stepwise process performed by (i) adjusting image contrast, (ii) utilizing a threshold tool to set lower and upper threshold values with the default method to segment the image into pore phase (black) and solid phase (white), and finally (iii) processing binary image with an open and then close transformations. The former procedure is performed as an erosion operation followed by a dilation, resulting in smoothing objects and removing isolated pixels. The latter procedure performed as a dilation followed by erosion, smoothes the objects and fills in small holes. The thresholded pore space, displayed in black as a foreground of binary image, was continuously measured in the course of image processing and the effects of the area change were checked.

Fractal dimension of pore space was determined using the box counting method with the *Standard Box Count Fractal Analysis* provided by a plugin *FracLac* for the ImageJ software [15]. This method is based on superimposing a series of grids of decreasing calibres (the box size, ε) onto an image and recording the number of boxes covering a pattern (the counting data, $N(\varepsilon)$), and enables to estimate a box counting fractal dimension (D_B) as a scaling rule from the relation between the number of pieces ($N(\varepsilon)$) and the scale (ε) used to get the new pieces:

$$N(\varepsilon) \propto \varepsilon^{-D_B} \tag{1}$$

A standard box count procedure with *FracLac* performed on the binary images of the pore space extracted as the foreground pixels involve setting up the options to make the set of grids used to sample an image (*Grid Design* panel). It was specified a power scaling method (a base of 2 and an exponent of 2) for calculating the series of decreasing grid calibers. The limits of box sizes were fixed between the smallest grid calibre of 2 pixels and the largest grid in a series equals to 45 % of the image width, whereas the number of grid calibres used in a series within these limits was determined by *FracLac* (setting a value of 0). Due to it ten grid orientations were set and the *FracLac* did multiple box counting scans. The first four positions of the grids are based on the four corners of an image, while all other positions of a grid are determined randomly around those four starting locations, within an area of the size of the biggest box in the series of grid calibres. Multiple box counting scans, delivering its own fractal dimension, allow to calculate the mean fractal dimension, the mean D_B, and the standard deviation $SD(D_B)$ [15].

2.3 Mercury Porosimetry and Fractal Dimension

The mercury intrusion porosimetry test consists in records of the incremental volumes of mercury intruding into a porous sample as the pressure is progressively increased, allowing time for equilibrium to be established at each pressure step. If a sample of a porous rock can be approximated by a system of cylindrical capillaries, then the application of the Washburn equation (2) to data from the mercury intrusion curve enables to evaluate the distribution of pore volume accessible by throats of a given effective diameter, *d*, corresponding to an intrusion pressure, *P*, viz:

$$d = -4\gamma \cos \theta / P \qquad (2)$$

where γ is the surface tension and θ is the contact angle of mercury [16, 17].

The pore volume distribution and, as has already been said, the apparent density and effective porosity were measured using a CE Instruments Pascal 140/240 high pressure mercury porosimeters that have a new pressurization system based on the PASCAL method (Pressurization by Automatic Speed-up and Continuous Adjustment Logic). The dilatometer containing each air-dried sample of diatomite was degassed under vacuum and then filled up with mercury. The pressure applied started at 0.145–0.395 MPa and ended at 150 MPa allows determination of pore throat diameters between 10.36–3.79 and 0.00998 μm. A value of the surface mercury tension of 0.480 N/m and a contact mercury angle on a siliceous rock of 141.3° were used in the Washburn equation (2), assuming cylindrical shapes of capillaries in the calculation. The mercury intrusion curves comprises 1615–1673 pairs of applied pressure and intruded volume values. The mercury porosimetry technique enables to determine the pore-throat size distribution in the macropore (pore width > 0.050 μm) and wide mesopore (mesopore width 0.002–0.050 μm) ranges as per the IUPAC terminology [17].

The assumption that the Menger sponge constructed from solid cubes of the density ρ_0 and the size r_0 is a model of a porous medium enables using a fractal relation between a density of a solid phase (ρ_n) and a size of a sample considered (r_n), namely:

$$\frac{\rho_n}{\rho_0} = \left(\frac{r_0}{r_n}\right)^{3-D} \quad (3)$$

where $D = \ln 20/\ln 3 = 2.727$ is the fractal dimension. Thus, for a fractal distribution of pore volume, the density of a rock (solid phase) decreases systematically with the increasing size of the sample considered [18]. With regard to the porosity of the Menger sponge, Φ, the relationship is expressed by the complement to 1 of the volume of the solid phase:

$$\Phi = 1 - \left(\frac{\rho_n}{\rho_0}\right) = 1 - \left(\frac{r_0}{r_n}\right)^{3-D} \quad (4)$$

This is not a power-law (fractal) relation due to porosity increases nonlinearly as a function of the sample size [19, 20].

The pore volume distribution of the Menger sponge may be measured by the mercury porosimetry technique. In the test considered, smaller and smaller pores are filled with mercury with increasing injection pressure so both the volume of mercury injected and the density of the sponge increase consequently. The total pore volume at a given pressure is given by the volume of mercury forced into a pore space network. Combining the equations (2) and (3) provides the following power-law relation between pore volume (V_P) and pressure (P):

$$V_P \propto P^{3-D_P} \quad (5)$$

where D_P is pore volume fractal dimension, which can be deduced from the slope of the linear log-log plot of pore volume, V_P, against injection pressure, P [21, 22].

3 Results and Discussion

3.1 Fractal Analysis of Pore Network from SEM Images

For all the ten SEM images binarized into pore and solid phases, the *total pore phase area* (*TPA*) measured in the calibrated unit (μm^2) was used to calculate the ratio of the *TPA* to the total image area, labelled as the *total optical porosity* (*TOP*) and expressed in percents [cf. 23] (Table 2). The values of the *TOP* of all the images range from 31.12 to 51.81 %, whereas the *TPA* varies from 17,803 to 33,405 μm^2 for the images of the larger pixel size (0.221–0.276 μm, i.e. the images of lower resolution and lower magnification) and from 664 to 914 μm^2 for the images of the smaller pixel size (0.021–0.024 μm, i.e. the images of higher resolution and higher magnification). The *TOP* values estimated for all the binary images are usually higher than the physically measured effective porosities of the samples (Table 1). Merely one *TOP* value (31.12 %) is consistent with the effective porosity of the sample (31.89 %).

The mean fractal dimension (mean D_B) calculated from multiple box counting scans of the pore phase extracted on the binary images, along with the standard deviation, $SD(D_B)$, is the measure of heterogeneity of the pore phase. However, different aspects of the pore network complexity become apparent depending on the level of image resolution [24]. The general features of the void system exhibited at low resolution (the images of low magnification) are quantitatively characterized by the *structural fractal dimension*; whereas the fine details of pore network revealed at high resolution (the images of high magnification) are described by the *textural fractal dimension.*

For the two resolutions the mean D_B of the pore phase varies from 1.7409 to 1.8436, and the values of standard deviation, $SD(D_B)$, are higher for the structural fractal dimensions (0.0343–0.0353), while for the textural fractal dimensions $SD(D_B)$ are lower (0.0189–0.0226) (Table 2). It has been found both for structural and textural fractals that the mean D_B is dependent on the *TOP* (%) and the *TPA* (μm^2). The coefficients of determination for the *TOP* (%) are equal to 0.9447 and 0.9984 (Fig. 1); while for the *TPA* (μm^2) 0.6176 and 0.7372, respectively (Fig. 2). Within the relatively narrow range of the values of the mean D_B (1.8012–1.8436) its

Table 2 Fractal analysis results of the pore phase of the Jawornik diatomites based on the SEM images at different magnifications (100-μm scale bar and 20-μm scale bar micrographs)

Sample	1	2	3	4	5
The SEM images of lower magnification (the structural fractal)					
Pixel size (μm)	0.2212	0.2757	0.2703	0.2212	0.2747
Total pore phase area (*TPA*) (μm^2)	20252.76	31272.12	33405.11	17803.22	28835.36
Total optical porosity (*TOP*) (%)	46.01	45.54	50.63	40.27	42.30
Mean fractal dimension (mean D_B)	1.8197	1.8280	1.8436	1.8023	1.8124
Standard deviation ($SD(D_B)$)	0.0353	0.0353	0.0352	0.0352	0.0343
The SEM images of higher magnification (the textural fractal)					
Pixel size (μm)	0.0228	0.0228	0.0209	0.0223	0.0243
Total pore phase area (*TPA*) (μm^2)	835.64	913.73	816.91	848.61	663.89
Total optical porosity (*TOP*) (%)	44.53	48.69	51.81	47.28	31.12
Mean fractal dimension (mean D_B)	1.8012	1.8164	1.8302	1.8093	1.7409
Standard deviation ($SD(D_B)$)	0.0226	0.0225	0.0222	0.0224	0.0189

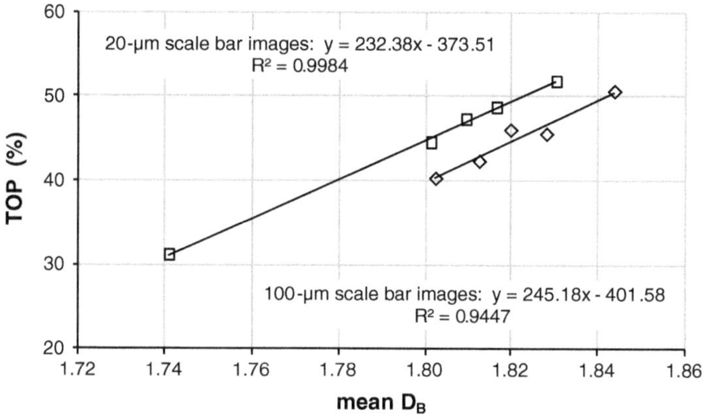

Fig. 1 The relations between the total optical porosity (*TOP*, %) and the mean box counting fractal dimension (mean D_B) evaluated for the structural and textural fractals revealed at low (*diamonds*) and high (*squares*) magnifications

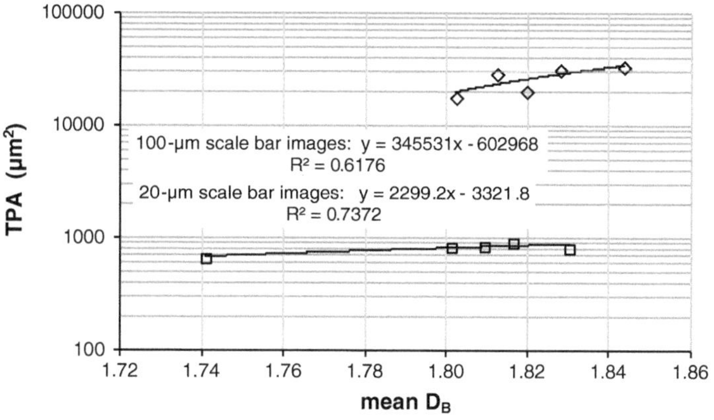

Fig. 2 The relations between the total pore phase area (*TPA*, µm^2) and the mean box counting fractal dimension (mean D_B) evaluated for the structural and textural fractals revealed at low (*diamonds*) and high (*squares*) magnifications

corresponding *TOP* values (44.53–51.81 %) are always higher for the textural fractals (Fig. 1).

Coincidentally, the selection of representative images for different magnifications indicated by scale bars affected the distribution of the pixel sizes. Therefore, the relation between the *TPA* (µm^2) and the pixel sizes (coefficient of determination $R^2 = 0.9013$) for the images of lower resolution has been found. Moreover, there exist the relations of the pixel sizes with the *TOP* ($R^2 = 0.8078$), as well as with the mean D_B ($R^2 = 0.8056$) for the images of higher resolution. Thus, the mean D_B of the latter images are more susceptible to the pixel size.

3.2 Fractal Analysis of Mercury Porosimetry Data

The calculations of the fractal dimensions of the void volume in diatomites based on the Menger sponge fractal model and using the mercury porosimetry data have been performed for the pore-throat diameters range from 10,000 to 10 nm.

The double-logarithmic intrusion curve of the cumulated void volume (log V_P), being accessible to mercury through pore throats plotted against the injection pressure (log P) for each sample reveals that the log V_P − log P relationship is linear merely in the region of the higher values of pressures (>22 MPa) corresponding to the smaller pore throat diameters (less than 70 nm). It has been found that there are two linear components within the above region of the log V_P – log P plot representing two separate fractal elements that have been revealed in two distinct ranges of the pressures. For each linear unit, fitting the line to the data by the least-squares linear regression was carried out for the as narrow as possible range of the cumulative pressures and for the as high as possible coefficient of determination (R^2) to estimate the fractal dimension.

The first fractal element tied to the range of the smallest pore throat diameters (10–20 nm) is characterized by the fractal dimensions (D_1) varying from 2.895 to 2.958, whereas the fractal dimensions (D_2) for the second fractal element linked to the range of the greater pore throat diameters (22.5–68.8 nm) extend from 2.712 to 2.846. The values of the coefficients of determination (R^2) are higher for the first fractal unit 0.9869–0.9946 and slightly lower for the second one 0.9821–0.9840 (Table 3).

Table 3 Fractal dimensions of the pore network of the Jawornik diatomites calculated from mercury porosimetry data

Sample	1	2	3	4	5
The first fractal element (the textural fractal)					
Fractal dimension (D_1)	2.8947	2.9577	2.9407	2.9342	2.9406
Coefficient of determination (R^2)	0.9929	0.9926	0.9869	0.9926	0.9946
Pore-throat diameter (nm)					
Min	9.976	9.976	9.976	9.976	9.976
Max	15.868	19.654	18.766	17.446	15.848
The second fractal element (the structural fractal)					
Fractal dimension (D_2)	2.8273	2.8459	2.7278	2.7798	2.7121
Coefficient of determination (R^2)	0.9821	0.9828	0.9827	0.9840	0.9829
Pore-throat diameter (nm)					
Min	28.472	30.470	27.914	22.500	27.626
Max	68.768	60.744	50.570	50.588	55.538
The inflection point of the rapidly rising portion of the injection curve					
Threshold pore-throat diameter (nm)	364.198	271.754	203.188	187.196	187.216
Threshold pressure (MPa)	4.114	5.514	7.375	8.005	8.004

Thus, the pore space volume distribution of diatomite samples exhibits the multifractal pattern that appears to have more than one fractal dimension at various levels of scrutiny. The first fractal, manifested at the range of the highest injection pressures and referred to as the *textural fractal*, can be related to the smallest nanoscale components of the pore system, becoming apparent at the highest level of scrutiny. On the other hand, the second fractal, exhibited at the region of lower injection pressures and described as the *structural fractal*, can be linked to the larger constituents of the pore network, being noticeable at a relatively low stage of scrutiny [24].

This study has revealed the existence of the multifractal pattern for the micro- and nanovolume distributions of the void space in diatomites, which is consistent with the hierarchical structure of nano and micro pore arrays of the diatom frustules that are a predominant constituent of diatomites [13]. Therefore it is valid to assume that the textural fractal dimension (D_1) characterizes the volume distribution of

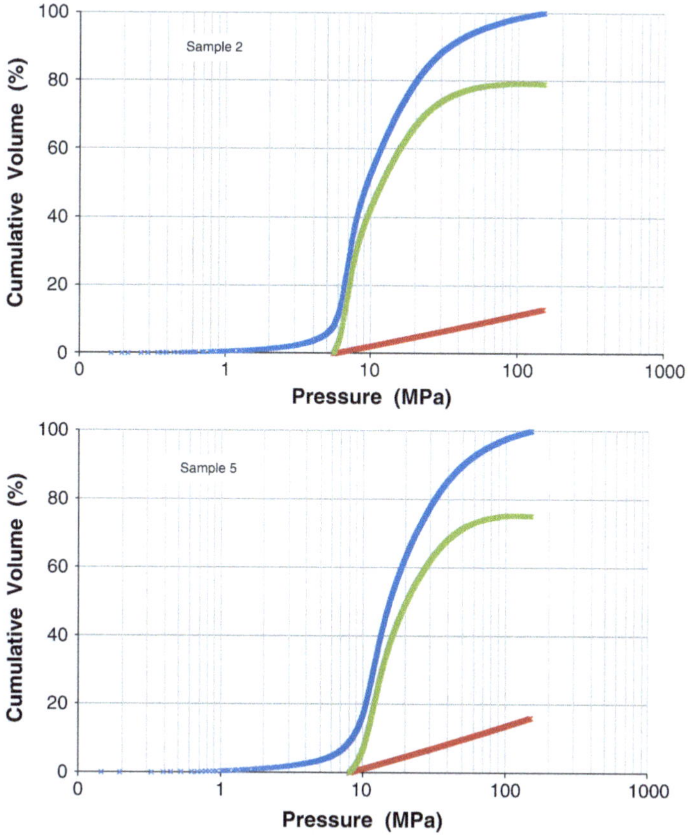

Fig. 3 The cumulative distributions of the pore-throat volume (*red*), the pore volume (*green*), and the pore-space volume (*blue*) for sample 2 and 5 of the Jawornik diatomites

nanoscale elements throughout the whole void system. This assumption yields a device whereby the cumulative distribution of the pore-space volume as a function of pressure can be partition into the pore-throat volume and the pore-body volume cumulative distributions. The lower values of the textural fractal dimension (D_1) result in the steeper distributions of the pore-throat volume, contrary to the more gradual distributions controlled by the higher values of D_1 (Fig. 3). This method of the partitioning the cumulative volume distributions introduced in studies of the pore space in sandstones [25, 26] has been applied in this work to investigate the pore system of diatomites.

4 Conclusions

The quantitative analysis of the pore network for a set of diatomite samples were carried out using the fractal approach to the mercury porosimetry data and the processed images of thin sections.

Both the box counting method implemented with a plugin *FracLac* for binary images of thin section micrographs taken at different magnifications, and the Menger sponge fractal model of a porous body applied for the mercury porosimetry data indicate that the void system of diatomites is a multifractal pattern. The fractal dimensions calculated using these two methodologies confirm that the pore space of diatomites is both a structural and textural fractal for different magnifications of 2-dimensional images as well as for different ranges of injection pressure. The multifractality of the pore network of diatomite samples is consistent with the micro- and nanoporous structure of diatom frustules. The fractal approach in study of the pore network for a set of diatomite samples, to our best knowledge, is applied for the first time.

Acknowledgments This work was financially supported by the AGH-UST statutory grant No. 11.11.140.320.

References

1. A.H. Thompson, A.J. Katz, C.E. Krohn, The microgeometry and transport properties of sedimentary rock. Adv. Phys. **36**(5), 625–694 (1987)
2. C.M. Ross, A.R. Kovscek, Pore microstructure and fluid distribution in a diatomaceous reservoir. SPE **75190**, 1–9 (2002)
3. L. Jia, C.M. Ross, A.R. Kovscek, A pore-network-modeling approach to predict petrophysical properties of diatomaceous reservoir rock. SPE **93806**, 597–608 (2007)
4. A.J. Katz, A.H. Thompson, Fractal sandstone pores: implications for conductivity and pore formation. Phys. Rev. Lett. **54**(12), 1325–1328 (1985)
5. C.E. Krohn, A.H. Thompson, Fractal sandstone pores: automated measurements using scanning-electron-microscope images. Phys. Rev. B **33**(9), 6366–6374 (1986)

6. J.P. Hansen, A.T. Skjeltorp, Fractal pore space and rock permeability implications. Phys. Rev. B **38**(4), 2635–2638 (1988)
7. C.E. Krohn, Sandstone fractal and Euclidean pore volume distributions. J. Geophys. Res. **93**(B4), 3286–3296 (1988)
8. C.E. Krohn, Fractal measurements of sandstones, shales and carbonates. J. Geophys. Res. **93**(B4), 3297–3305 (1988)
9. B.B. Mandelbrot, *The Fractal Geometry of Nature* (Freeman, New York, 1983)
10. P. Pfeifer, D. Avnir, Chemistry in noninteger dimensions between two and three. I. Fractal theory of heterogeneous surfaces. J. Chem. Phys. **79**(7), 3558–3565 (1983)
11. J. Kotlarczyk, Leszczawka Diatomite Horizon in diatomite quarry near Jaworowice (former Jawornik Ruski), in *Geotraverse Kraków-Baranów-Rzeszów-Przemyśl-Ustrzyki Dolne-Komańcza-Dukla*, Guide to Excursion 4, Carpatho-Balkan Geological Association, XIII Congress, Cracow, Poland, ed. by J. Kotlarczyk (Geological Institute, Warszawa, 1985), pp. 145–147
12. S. Boggs Jr., *Petrology of Sedimentary Rocks* (Cambridge University Press, Cambridge, 2009)
13. W. Yang, P.J. Lopez, G. Rosengarten, Diatoms: self assembled silica nanostructures, and templates for bio/chemical sensors and biomimetic membranes. Analyst **136**, 42–53 (2011)
14. W. Rasband, ImageJ, 1.42 g ed. National Institutes of Health, USA Java 1.6.0_20 (64-bit). http://rsbweb.nih.gov/ij
15. A. Karperien, FracLac for ImageJ http://rsb.info.nih.gov/ij/plugins/fraclac/FLHelp/ (1999–2013)
16. C.L. Vavra, J.G. Kaldi, R.M. Sneider, Geological applications of capillary pressure: a review. AAPG Bull. **76**(6), 840–850 (1992)
17. IUPAC, Manual of symbols and terminology for physicochemical quantities and units—appendix II. Definitions, terminology and symbols in colloid and surface chemistry. Part II: heterogeneous catalysis. Pure Appl. Chem. **46**(1), 71–90 (1976)
18. D.L. Turcotte, J. Huang, Fractal distribution in geology, scale invariance, and deterministic chaos, in *Fractals in the Earth Sciences*, ed. by C.C. Barton, P.R. La Pointe (Plenum Press, New York, 1995), pp. 1–40
19. D.L. Turcotte, *Fractals and Chaos in Geology and Geophysics* (Cambridge University Press, Cambridge, 1992)
20. F. Bartoli, N.R.A. Bird, V. Gomendy, H. Vivier, S. Niquet, The relation between silty soil structures and their mercury porosimetry curve counterparts: fractals and percolation. Eur. J. Soil Sci. **50**, 9–22 (1999)
21. R.F. Angulo, V. Alvarado, H. Gonzalez, Fractal dimensions from mercury intrusion capillary tests. SPE **23695**, 255–263 (1992)
22. J.D. Bonny, H. Leuenberger, Determination of fractal dimensions of matrix-type solid dosage forms and their relation with drug dissolution kinetics. Eur. J. Pharm. Biopharm. **39**(1), 31–37 (1993)
23. R. Ehrlich, S.J. Crabtree, K.O. Horkowitz, J.P. Horkowitz, Petrography and reservoir physics I: objective classification of reservoir porosity. AAPG Bull. **75**(10), 1547–1562 (1991)
24. B.H. Kaye, Image analysis techniques for characterizing fractal structures, in *The Fractal Approach to Heterogeneous Chemistry*, ed. by D. Avnir (Wiley, Chichester, 1989), pp. 55–66
25. P. Such, G. Lesniak, Study of pore space parameters of rocks. Prace Instytutu Gornictwa Naftowego i Gazownictwa **119**, 3–63 (2003) (summary in English)
26. G. Stanczak, Fractal analysis of the pore space in sandstones as derived from mercury porosimetry and image analysis, in *International Multidisciplinary Microscopy Congress*, ed. by E.K. Polychroniadis, A.Y. Oral, M. Ozer (Springer Proceedings in Physics 154, 2014) pp. 51–60

Development of an Off-Axis Digital Holographic Microscope for Large Scale Measurement in Fluid Mechanics

K.F. Tamrin, B. Rahmatullah and S.M. Samuri

Abstract Holographic particle image velocimetry is a promising technique to probe and characterize complex flow dynamics since it is a truly three-dimensional (3D) three-component measurement technique. The technique simply records the coherent light scattered by small seeding particles that are assumed to faithfully follow the flow and uses it to reconstruct the event afterward. Reconstruction of the event is usually performed using a digital video microscope mounted on a 3D translation stage. The microscope records the intensity only which consequently results in loss of phase information. The objective of this paper is to develop and apply digital holographic microscopy with the aim to recover the phase information. Digital holographic microscopy has immense potentials in microscale solid and fluid measurements as it offers the possibility of digital wavefront processing by manipulating amplitude and phase of the recorded holograms. In this paper, we have developed an off-axis digital holographic microscope to capture both amplitude and phase of the reconstructed object simultaneously. This inherently solves twin image problem in the recorded digital holograms. The microscope was integrated into the reconstruction system and was successfully used to digitize holographic images of 10 μm polystyrene spheres and 1 μm olive oil droplets. The spatial resolution of the system is 0.63 μm, and the field of view is 1250 × 625 μm^2. A 3D holographic reconstruction using a k-space analysis (wave-vector) of the optical field is applied to numerically refocus the images. Another potential application includes digital wavefront processing to compensate for aberration in the images.

K.F. Tamrin (✉) · B. Rahmatullah · S.M. Samuri
Computing Department, Faculty of Arts, Computing and Creative Industry, Universiti Pendidikan Sultan Idris, Tanjong Malim, 35900 Perak, Malaysia
e-mail: k.f.tamrin@outlook.com

E.K. Polychroniadis et al. (eds.), *2nd International Multidisciplinary Microscopy and Microanalysis Congress*, Springer Proceedings in Physics 164,
DOI 10.1007/978-3-319-16919-4_12

1 Introduction

Holographic particle image velocimetry (HPIV) is a promising technique to probe and characterize complex flow dynamics since it is a truly three-dimensional (3D) three-component measurement technique. The technique relies on recording the coherent light scattered by small seeding particles that are assumed to faithfully follow the flow and uses it to reconstruct the event afterward. The essence of holographic imaging lies in the fact that, when coherent light propagates through a semi-transparent object, its amplitude and phase get modulated due to light-matter interaction. This effectively means that the entire 3D structure of the object is coded in the form of scattered wavefronts which eventually incident on to holographic films [1]. However, reconstruction of the event is usually performed using a digital video microscope mounted on a 3D translation stage [2]. The microscope records the intensity only which consequently results in loss of phase information.

On the other hand, a new application of holography in the field of optical microscopy has introduced a new imaging technology, known as digital holographic microscopy [3]. It is essentially similar to classical holography except that CCD sensors are used instead of holographic film emulsion [4]. For this reason, one can directly record information about the phase and amplitude of the object's wavefront and thus numerically reconstruct sectional images of specimens at different depths from only a single digital hologram [3]. The remarkable aspect of digital reconstruction is the topic of the present study. Digital holographic microscopy has proved to be useful in solid and fluid measurements since it allows digital processing of the wavefronts [5, 6]. In comparison to HPIV, the measurement volume with digital holographic microscopy is significantly less due to inferior pixel resolution of the imaging sensor.

In this paper, an off-axis digital holographic microscope was developed and integrated in the hologram reconstruction system to capture both amplitude and phase of the reconstructed object simultaneously. By combining the power of optics with digital holography, one is able to perform fluid measurements in a bigger volume with microns resolution. In the following section, the construction of the microscope is first discussed.

2 Digital Holographic Microscope

Unlike typical HPIV reconstruction approach which utilised a microscope objective [2], the novel approach of this study involves the employment of a digital holographic microscope (DHM) that allows simultaneous digitisation of holographic particle image in both amplitude and phase formats. The schematic of the DHM assembly is illustrated in Fig. 1.

The reconstructed image is first magnified using a 10× microscope objective. To compensate for phase curvature [7] introduced by the microscope objective to the

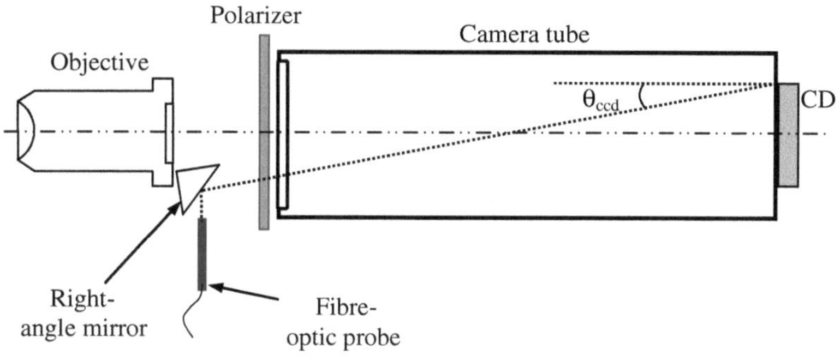

Fig. 1 Schematic diagram of a digital off-axis holographic microscope. The magnified image of the reconstructed object and the reference wave interferes on the CCD plane with an intersecting angle, θ_{ccd} less than that of Nyquist's limit. An off-axis digital hologram of the object is finally recorded

object wave, the reference wave of an approximately the same curvature was introduced at the fibre tip with numerical aperture of about 0.10–0.14. The magnified image of the reconstructed object and the reference wave interferes on the CCD plane with an intersecting angle, θ_{ccd} less than that of Nyquist's limit. The interference patterns are finally recorded on a Lumenera 10.8 MPixel digital camera with pixel size of 9 μm^2.

An off-axis digital hologram of the object is finally recorded. It is found that the spatial resolution of the system is 0.63 μm, and the field of view is 1250 × 625 μm^2. The DHM is then mounted on a 3D-motorized stage, permitting detailed measurement in a much bigger imaging volume (>100^3 mm^3) with micrometres resolution. Similar principles of digital holographic recording and reconstruction that have been described in [8] are utilised to process the digital holograms. The developed DHM is tested in the following section.

3 Experiment

To verify the applicability of the DHM for large scale measurement, two different particle sizes were tested: (a) 10 ± 0.05 μm NIST traceable polystyrene microspheres of density 1.05 g/cm^3 (Thermo Fisher Scientific) in aqueous form, and (b) 1 ± 0.3 μm produced using a TSI six-jet atomiser (Model 9306) that was set to run on two jets at about 100 kPa. These objects were recorded on two separate holograms.

These objects were holographically recorded using an off-axis recording setup. A frequency doubled Nd:YAG laser operating at 532 nm (Continuum Inc.) was used for the illumination and the beam energy was equally split into the respective object and reference paths. The laser was set producing a single pulse of 1.3 ns width with energy of 84 $\mu J/cm^2$. A collimated reference beam was carefully

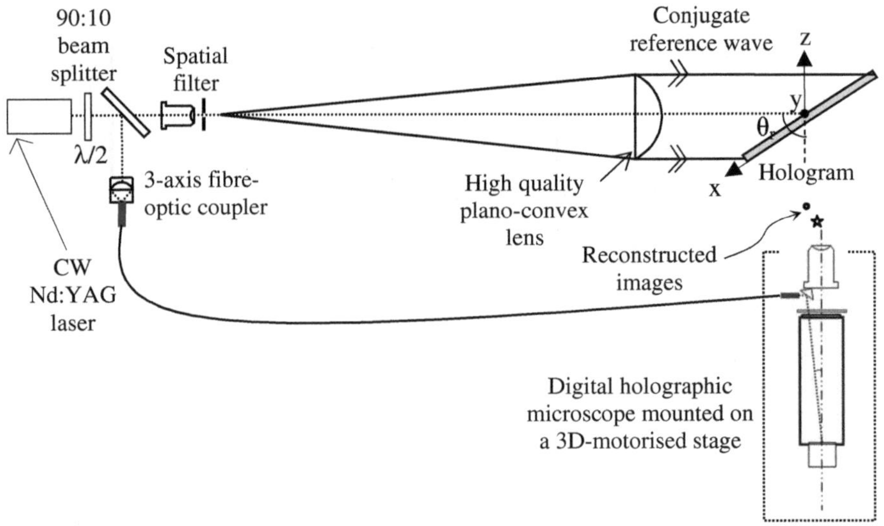

Fig. 2 Hologram reconstruction setup and utilisation of an off-axis digital holographic microscope to digitise both the amplitude and phase of the reconstructed objects. A conjugate reference wave is initially employed to reconstruct the hologram, and eventually the reconstructed images are digitised using the DHM

produced which incident on the plane of the hologram at Brewster's angle whilst a diverging object beam was used to illuminate the objects. The interference was recorded on a Slavich VRP-M holographic plate of Russian origin (having resolving power of 3000 lines/mm). The hologram was developed in Kodak D-19 for about 3 min maintained at ~25 °C during the development process. There was no fixing or bleaching step adopted. Once the hologram was developed, the reconstructed image of the (aberrated) objects was digitised using the purpose built digital holographic microscope as follows.

An experimental setup as shown in Fig. 2 was used to reconstruct the off-axis optical hologram. The first step involved an optical reconstruction of the hologram using a phase-conjugate of the reference wave, producing real image of the objects. In this setup, a single longitudinal mode (CNI MSL-III-532 laser, 532 nm, 80 mW) laser with long coherence length of about 50 m was used and split by a 90:10 plate beam splitter, where most of the power is directed towards reconstructing the hologram. The remaining power was transmitted via an optical fibre to form an off-axis reference wave. By referring to Fig. 2, the intensity of the reference wave was adjusted using a variable neutral density filter and its polarization angle was compensated by a linear polarizer. Thereafter, the DHM was used to scan in a relatively large three-dimensional space and simultaneously digitise both amplitude and phase information of the images.

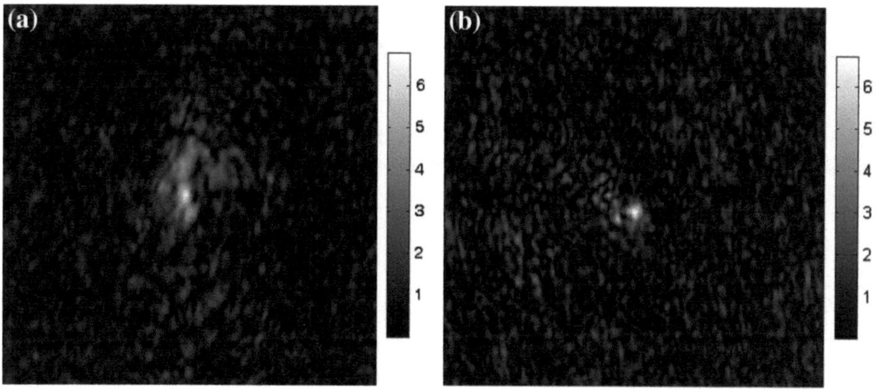

Fig. 3 Demodulated amplitude images of **a** 10 μm polystyrene sphere and **b** 1 μm olive oil droplet. Each image size is 236 × 236 μm^2. Intensity equals to the square of the amplitude

4 Results and Discussion

The developed DHM was used to capture both amplitude and phase of the reconstructed 10 μm polystyrene sphere and 1 μm olive oil droplet from two separate optical holograms. A 3D holographic reconstruction using a k-space analysis (wave-vector) of the optical field is applied to numerically refocus the images [8]. This indirectly overcomes limited resolution along the z-axis imposed by the 3D mechanical translation state. The demodulated amplitude images after numerical refocus are shown in Fig. 3.

Further examination reveals that the images appear to be influenced by aberrations. It can be seen that as the particle size gets smaller the severity of aberration is less pronounced. It is noted that aberration severely degrades the image quality and this eventually would make particle centroid identification problematic. Compensation of the aberration in the particle images will be the subject of our future study.

5 Conclusions

Holographic particle image velocimetry is truly the key for three-dimensional three-component fluid flow measurement over a deep measuring volume. Digitisation of the reconstructed images using a video microscope however results in loss of useful phase information. In this paper, a new off-axis digital holographic microscope has been developed with the aim to recover the loss. The developed microscope was successfully used to digitize holographic images of 10 μm polystyrene spheres and 1 μm olive oil droplets. The spatial resolution of the system is 0.63 μm, and the field of view is 1250 × 625 μm^2. We have demonstrated the benefits of having both phase and amplitude for numerical refocus of the object. The DHM has the potential for digital wavefront processing to compensate for aberration found in the images.

Acknowledgments The authors are grateful to Prof. J.M. Coupland (Loughborough University) for useful discussion and access given to perform the experiments.

References

1. D. Gabor, A new microscopic principle. Nature **161**(4098), 777–778 (1948)
2. K.D. Hinsch, Holographic particle image velocimetry. Meas. Sci. Technol. **13**(7), R61 (2002)
3. U. Schnars, W. Jueptner, Digital holographic microscopy. Digital holography: digital hologram recording, numerical reconstruction, and related techniques, pp. 95–99 (2005)
4. H. Meng, G. Pan, Y. Pu, S.H. Woodward, Holographic particle image velocimetry: from film to digital recording. Meas. Sci. Technol. **15**(4), 673 (2004)
5. T. Colomb, E. Cuche, F. Charrière, J. Kühn, N. Aspert, F. Montfort, P. Marquet, C. Depeursinge, Automatic procedure for aberration compensation in digital holographic microscopy and applications to specimen shape compensation. Appl. Opt. **45**(5), 851–863 (2006)
6. S. Satake, T. Kunugi, K. Sato, T. Ito, H. Kanamori, J. Taniguchi, Measurements of 3D flow in a micro-pipe via micro digital holographic particle tracking velocimetry. Meas. Sci. Technol. **17** (7), 1647 (2006)
7. J.W. Goodman, Introduction to Fourier Optics. Roberts & Company Publishers, Englewood (1996)
8. S.A. Wormald, J. Coupland, Particle image identification and correlation analysis in microscopic holographic particle image velocimetry. Appl. Opt. **48**(33), 6400–6407 (2009)

SEM-EDS Observation of Structure Changes in Synthetic Zeolites Modified for CO_2 Capture Needs

Magdalena Wdowin, Rafal Panek and Wojciech Franus

Abstract Carbon dioxide is the main greenhouse gas and its amount still increase in the atmosphere. Air pollution and greenhouse effect caused by CO_2 emission have become a major threat to the environment on a global scale. Carbon dioxide sequestration (i.e. capture and consequently geological storage) is the key strategy within the portfolio of actions to reduce CO_2 emission to the atmosphere. The most costly stage is capture of CO_2, therefore there is a need to search new solutions of this technology. For this purpose it was examined Na–X synthetic zeolites, that were silver and PEI (polyethyleneimine) activated. SEM-EDS investigation enable to find a changes in structure of this materials after treatment. Where, as a result of silver activation from EDS analysis it is seen that Ag occur in Na–X structure, what indicate a substitution of Ag^{2+} for Na^{+} ions in crystal lattice. Analysing wt% the EDS analysis has shown that zeolite Na–X after silver impregnation becomes Ag–X zeolite. For Na–X–PEI activated it is observed a distinct organic compound in the form of coatings on Na–X crystals causing a sealing of pores in tested zeolite. Further examination of these materials concern determination of surface properties and experiments of CO_2 sorption. But SEM-EDS analysis enable to determine the extent of activation, what is very important in determination of optimal conditions for such treatment in order to obtain better sorbent of CO_2.

M. Wdowin (✉)
The Mineral and Energy Economy Research Institute of the Polish Academy of Sciences, Wybickiego 7, 31-261 Kraków, Poland
e-mail: wdowin@meeri.pl

R. Panek · W. Franus
Division of Geotechnics, Lublin University of Technology, Nadbystrzycka 40, 20-618 Lublin, Poland
e-mail: r.panek@pollub.pl

W. Franus
e-mail: w.franus@pollub.pl

E.K. Polychroniadis et al. (eds.), *2nd International Multidisciplinary Microscopy and Microanalysis Congress*, Springer Proceedings in Physics 164,
DOI 10.1007/978-3-319-16919-4_13

1 Introduction

Combustion and energy production in Poland mainly is based on coal (bituminous and lignite), which is related to high CO_2 emissions to the atmosphere [1]. This production is also associated with the production of wastes such as fly ash, that are dangerous for environment and the human health, mainly due to leaching of heavy metals into the soil and release of light fractions into the atmosphere [2–4].

One of the latest solutions of fly ash utilization is its usage in the synthesis of zeolites [5, 6], which (because of its unique properties) are used in many fields of environmental engineering and protection among others refining of crude oil [7], the removal of ammonium ions from wastewater [8, 9] and heavy metals [10] or radionuclides from mine waters [11] but also in the separation/adsorption of gases, such as CO_2 [12, 13], SO_2 [14] or gaseous forms of mercury [15, 16].

One of the ways to solve the problem carbon dioxide emissions is the technology of Carbon Capture and Storage (CCS), involving the capture of CO_2 from point sources of emissions, followed by its transport and storage in the suitable geological structures [17], that often constituted hydrogeothermal structures abundant in geothermal water exploited for the direct management in heating or balneology as well as recreation [18, 19]. At large extent CCS studies are focused on the search for new capture technology due to the fact that this process is 84 % of the total cost of CCS [20]. A number of adsorbents are currently under development and some were demonstrated at small pilot scale, including supported and immobilised amines, functionalised carbon materials and zeolites [21].

Zeolites, molecular sieves, as well as activated carbon have been conventionally used for CO_2 capture and separation by selective adsorption. However, adsorption capacity for CO_2 of these materials decreases drastically with increased temperature. Furthermore, zeolites have very low CO_2 adsorption capacities in the presence of moisture that is present in flue gas and high uptakes are thus only obtained under dry conditions [22]. In order to enhanced their capacities terms to carbon dioxide it is necessary to impregnated various compounds among others: amine [23, 24] as well as ion substitutions [12, 25] etc.

One of the basic methods for fast assessment of the impregnation extent of tested zeolite materials is usage for this purpose the scanning electron microscopy. In the case of ion substitutions the SEM-EDS analysis is very suitable. The extent of impregnation of organic compounds can be precisely determined by observing the nature and morphology of the crystal grains of tested materials [26].

The aim of the study was to determine the degree of Na–X zeolite impregnation by silver ions and polyethylenimine in order to use it as a sorbent of carbon dioxide.

2 Materials

The research material was synthetic zeolite Na–X type, that represents a group of faujasite, according to the classification based on secondary building units (SBU) belonging to the D6R group. This zeolite was obtained in the hydrothermal reaction of fly ash with aqua solution of NaOH according to the following formula [5, 6]:

$$\text{Fly ash} + x * \text{mol} * \text{dm}^{-3}\text{NaOH} \ \overrightarrow{\text{T, h}} \ \text{Zeolite} + \text{residuum}$$

where the following process conditions were applied: 20 g of fly ash was mixed with 0.5 dm^3 of NaOH at a concentration of 3 mol dm^{-3} for 24 h at 75 °C.

In order to improve the adsorption capacity of the tested zeolite for CO_2, it was activated in two ways: by ion exchange (exchanging sodium ions by silver ions) and impregnated with organic compounds of polyethyleneimine.

Silver activation was carried out according to the modified method proposed by Stein et al. [27], where 100 g of zeolite was added to 500 mL of 0.5 mol dm^{-3} of $AgNO_3$ solution and then shaken (speed of 180 times/min) in the dark for 24 h at room temperature. Then the zeolite was dried in air-dry state.

Activation of polyethyleneimine (PEI) was performed according to modified method proposed by Xu et al. [28], wherein 25 g of PEI was mixed with 100 g of methanol and then shaken for 15 min. Next, 25 g of analyzed zeolite was added and shaken for 30 min. Subsequently the samples were dried at 100 °C for 24 h.

3 Methods

The morphological forms and the chemical composition were determined by the means of a scanning electron microscope (SEM) FEI Quanta 250 FEG equipped with a system of chemical composition analysis based on energy dispersive X-ray-EDS of EDAX company [16].

As a complementary study to determine the mineral composition and the degree of crystallization of the tested materials XRD diffraction studies was performed.

The XRD analysis was carried out via powder X-ray diffraction (XRD) method using a PANalytical X'pert APD diffractometer (with a PW 3020 goniometer), Cu lamp, and a graphite monochromator. The measurements were done within the angle range of 5–65 2θ. Philips X'Pert Highscore software was used to process the diffraction data as well as the identification of mineral phases was based on the PDF-2 release 2010 database formalized by the ICDD [16].

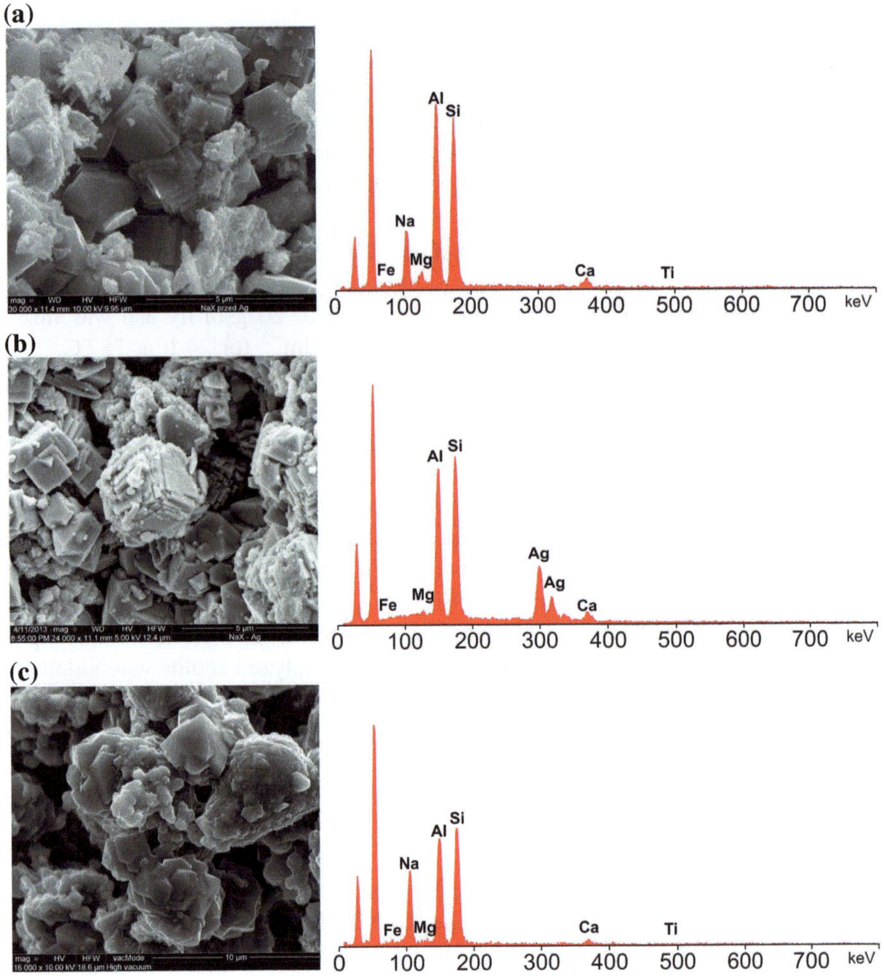

Fig. 1 SEM-EDS analysis of tested zeolite materials, **a** Na–X; **b** Ag–X, **c** PEI–Na–X

4 Results

Carried out observations of grain morphology for both Na–X, Ag–X and PEI–Na–X zeolites have shown isometric (cubic) and octahedral forms (Fig. 1).

Zeolite crystal sizes were from 2 to 6 microns. Whereas from the SEM-EDS analysis average ratio of the sum of the individual cations (which balancing the charge of the aluminosilicate lattice) were calculated. The ratios were as follow: for Na–X $Na_2O + K_2O/SiO_2 = 0.15$; $Na_2O + K_2O + CaO + MgO/SiO_2 = 0.31$; $SiO_2/Al_2O = 0.96$ and for zeolite Ag–X $AgO + K_2O/SiO_2 = 1.55$; $AgO + K_2O + CaO + MgO/SiO_2 = 1.61$; $SiO_2/Al_2O = 1.08$. Chemical analyzes (SEM-EDS) performed in the microarea confirmed complete replacement of Na^+ ions by Ag^+ ions.

In addition, SEM-EDS studies have shown that activation of the amine compound polyethyleneimine (PEI) caused blockage/sealing of the pore space in the studied zeolite material, and thus the accumulation of amine compounds on the surface of the tested material. In the SEM microphotographs (Fig. 1) it is apparent that the compound of PEI formed on zeolite coatings which constitute a seal for easy entering into the structure of zeolite moisture.

The XRD results showed that Na–X zeolite is the predominant product of the conversion reaction of fly ash. Its presence was determined on the basis of the characteristic main d-spacing (d_{hkl} = 14.47, 3.81, 5.73, 8.85, 4.42, 7.54, 4.81, 3.94 Å). Activation of the structure of Na–X zeolite by silver ions in the diffraction patterns is observed by a slight shift of the main peaks of the basic phase and the disappearance of reflections about the distances d = 8.93 and d = 5.76 Å. The activation did not affect the intensity of individual reflections originated from Na–X zeolite.

In case of PEI activation is it observe only higher background in the XRD diffraction curve in comparison to XRD of initial Na–X zeolite.

5 Conclusions

The presented results of microscopic observation and SEM-EDS analysis indicate that the method of scanning electron microscopy can be a basic method in determining the extent of zeolite activations/impregnations for use them as CO_2 sorbents.

Chemical analysis in the microarea (SEM-EDS) determines with high accuracy the extent of ion substitutions in the activation of Na–X zeolite by silver ions.

Observations of grain morphology enable to determine the extent of impregnation of the polyethyleneimine compound (the more visible shape of grain morphology the lower the degree of impregnation of the zeolite [26]).

Acknowledgments This research was financed by NCBiR within the Project No IPBU.01.01.00-06-570/11-00 and No POKL.04.03.00-00-129/12 and No NCBR/FENCO-NET 1/2013.

References

1. M. Wdowin, R. Panek, W. Franus, Badania właściwości zeolitów otrzymywanych z popiołów lotnych pod kątem wykorzystania ich jako sorbentów CO_2. Polityka Energetyczna **17**(4), 329–338 (2014)
2. M. Chudek, J. Hycnar, S. Janiczek, F. Plewa, *Węgiel brunatny* (Utylizacja surowców towarzyszących i odpadów elektrownianych. Wydawnictwo Politechniki Śląskiej, Gliwice, 1999)
3. J.C. Swanepoel, C.A. Strydom, Utilisation of fly ash in a geopolymeric material. Appl. Geochem. **17**, 1143–1148 (2002)

4. W. Franus, M. Wdowin, Wykorzystanie popiołów lotnych klasy F do produkcji materiału zeolitowego na skalę półtechniczną. Polityka Energetyczna **14**(2), 79–91 (2011)
5. W. Franus, Characterization of X-type zeolite prepared from coal fly ash. Pol. J. Environ. Stud. **21**(2), 337–343 (2012)
6. W. Franus, M. Wdowin, M. Franus, Synthesis of zeolites for fly ash development. Environ. Monit. Assess. **186**(9), 5721–5729 (2014)
7. J. Zhu, X. Meng, F. Xiao, Mesoporous zeolites as efficient catalysts for oil refining and natural gas conversion. Front. Chem. Sci. Eng. **7**(2), 233–248 (2013)
8. W. Franus, M. Wdowin, Removal of ammonium ions by selected natural and synthetic zeolites. Gosp. Sur. Miner.-Miner. Resour. Manag. **26**(4), 133–148 (2010)
9. H. Liu, S. Peng, L. Shu, T. Chen, T. Bao, R.L. Frost, Effect of Fe_3O_4 addition on removal of ammonium by zeolite NaA. J. Colloid Interface Sci. **390**(1), 204–210 (2012)
10. H. Merrikhpour, M. Jalali, Comparative and competitive adsorption of cadmium, copper, nickel, and lead ions by Iranian natural zeolite. Clean Techn. Environ. Policy **15**, 303–316 (2013)
11. S. Chałupnik, W. Franus, M. Wysocka, G. Gzyl, Application of zeolites for radium removal from mine water. Environ. Sci. Pollut. Res. **20**(11), 7900–7906 (2013)
12. K.S. Walton, M.B. Abney, L.M. Douglas, CO_2 adsorption in Y and X zeolites modified by alkali metal cation exchange. Micropor. Mesopor. Mater. **91**, 78–84 (2006)
13. M. Wdowin, W. Franus, R. Panek, Preliminary results of usage possibilities of carbonate and zeolitic sorbents in CO_2 capture. Fresen. Environ. Bull **21**(12), 3726–3734 (2012)
14. H. Yi, H. Deng, X. Tang, Q. Yu, X. Zhou, H. Liu, Adsorption equilibrium and kinetics for SO_2, NO, CO_2 on zeolites FAU and LTA. J. Hazard. Mater. **15**(203–204), 111–117 (2012)
15. J.R. Morency, T. Panagiotou, C.L. in *Senior, Laboratory Duct Injection of a Zeolite-Based Mercury Sorbent. Presented at the Annual Meeting of the Air and Waste Management Association*, Salt Lake City, UT, June 2000
16. M. Wdowin, M.M. Wiatros-Motyka, R. Panek, L.A. Stevens, W. Franus, C.E. Snape, Experimental study of mercury removal from exhaust gases. Fuel **128**, 451–457 (2014)
17. P. Baran, K. Zarębska, A. Nodzeński, Energy aspects of CO_2 sorption in the context of sequestration in coal deposits. J. Earth Sci. **25**(4), 719–726 (2014)
18. B. Tomaszewska, L. Pająk, Geothermal water resources management—economic aspects of their treatment. Gosp. Sur. Miner.-Miner. Resour. Manag. **28**(4), 59–70 (2012)
19. B. Tomaszewska, M. Bodzek, The removal of radionuclides during desalination of geothermal waters containing boron using the BWRO system. Desalination **309**, 284–290 (2013)
20. IPCC Special Report on Carbon Dioxide Capture and Storage. Prepared by Working Group III of the Intergovernmental Panel on Climate Change, ed. by B. Metz, O. Davidson, H.C. de Coninck, M. Loos, L.A. Meyer. (Cambridge University Press, Cambridge, New York, 2005)
21. S. Choi, J.H. Drese, C.W. Jones, Adsorbent materials for carbon dioxide capture from large anthropogenic point sources. Chemsuschem. **2**(9), 796–854 (2009)
22. F. Su, C. Lu, CO_2 capture from gas stream by zeolite 12X using a dual-column temperature/vacuum swing adsorption. Energy Environ. Sci. **5**, 9021–9027 (2012)
23. R. Idem, M. Wilson, P. Tontiwachwuthikul, A. Chakma, A. Vewab, A. Aroonwilas, D. Gelowitz, Pilot plant studies of the CO_2 capture performance of aqueous MEA and mixed MEA/MDEA solvents at university of regina CO_2 capture technology development plant and the boundary dam CO_2 demonstration plant. In. Eng. Chem. Res. **45**(8), 2414–2420 (2006)
24. G. Qi, Y. Wang, L. Estevez, X. Duan, N. Anako, A.H.A. Park, W. Li, C.W. Jones, E.P. Giannelis, High efficiency nanocomposite sorbents for CO_2 capture based on amine-functionalized mesoporous capsules. Energy Environ. Sci. **4**, 444–452 (2011)
25. A. Khelifa, L. Benchehida, Z. Derriche, Adsorption of carbon dioxide by X zeolites exchanged with Ni^{2+} and Cr^{3+}: isotherms and isosteric heat. J. Colloid Interf. Sci. **278**, 9–17 (2004)
26. J. Wang, H. Chen, H. Zhou, X. Liu, W. Qiao, D. Long, L. Ling, Carbon dioxide capture using polyethylenimine-loaded mesoporous carbons. J. Envi. Sci. **25**(1), 124–132 (2013)

27. A. Stein, G.E. Ozin, P.M. Macdonald, G.D. Stucky, R. Jelinek, Class A sodalites: silver, sodium halosodalites. J. Am. Chem. Soc. **114**(3), 5171–5186 (1992)
28. X. Xu, C. Song, J.M. Andrésen, B.G. Miller, A.W. Scaroni, Preparation and characterization of novel molecular basket adsorbent based on polymer-modified mesoporous molecular sieve MCM-41. Micropor. Mesopor. Mater. **62**, 29–45 (2003)

SEM Investigation of Microstructures in Hydration Products of Portland Cement

Wojciech Franus, Rafal Panek and Magdalena Wdowin

Abstract Portland cement is the one of the main ingredients in the manufacture of many building materials include concrete composites. The phase composition of hydration products is controlled by means of XRD and DTA/TG analysis. Observations of phase changes and microstructure of maturing cement pastes can also be observed using scanning electron microscopy (SEM) techniques combined with chemical analyzes in microareas (EDS analysis). Hydrating in time cement paste is composed mainly of hydrated silicates of calcium so called C–S–H-phase accompanied with a calcium hydroxide (portlandite) and hydration products of calcium aluminate i.e. ettringite. SEM analysis of changes in the morphology and microstructure of cement pastes allow to track the hydration progress observed mainly by changes in the C–S–H phase. The initial stages of hydration of this phase is characterized by radial concentration of fibers or needles, often narrowed at the ends. This fibers grow from the surface of the cement grains. The increase in the degree of C–S–H structure orientation is shown by formation of fibers lattice, sometimes three-dimensional plates so called "honeycomb", which is transformed into the form of a closely-packed, isometric grains. In addition, besides C–S–H cement phase investigation, scanning electron microscopy can also be applied to observation of the crystals formation of tobermorite, ettringite and relicts of portlandite, that often can't be detected by XRD and DTA/TG due to their small amount in mineral composition of concrete.

W. Franus (✉) · R. Panek
Division of Geotechnics, Lublin University of Technology, Nadbystrzycka 40, 20-618 Lublin, Poland
e-mail: w.franus@pollub.pl

R. Panek
e-mail: r.panek@pollub.pl

M. Wdowin
Mineral and Energy Economy Research Institute of the Polish Academy of Sciences, Wybickiego 7, 31-261 Kraków, Poland
e-mail: wdowin@meeri.pl

E.K. Polychroniadis et al. (eds.), *2nd International Multidisciplinary Microscopy and Microanalysis Congress*, Springer Proceedings in Physics 164,
DOI 10.1007/978-3-319-16919-4_14

1 Introduction

Knowledge of the basic chemical reactions and physical phenomena as well as microstructures of the hardened mass formed by the hydration and hydrolysis of cement is one of the basic conditions for the optimal use of cement in building materials. Predicting the cement properties on the basis of its chemical and mineral composition is essential for its use [1]. No fully define mechanism of the cement hydration and hydrolysis processes does not allow for explanation anomalies in the process of binding and hardening. Strength of hardened cement paste is a function of: mineral and chemical composition as well as has a direct impact on the structure that determines the strength of the crystalline lattice and tightness of mass. Therefore, they are needed studies of crystallizing mineral phases changes (hydration products) and observations of the reconstruction of the microstructure of maturing cement paste [2]. Method used to determine this type of phenomena may be scanning electron microscopy combined with the chemical composition analysis (SEM-EDS).

Quantitatively, the main component of the most cement paste and at the same time the most important factor affecting the strength of the hardening mass is phase denoted by the symbol C–S–H—often called phase of tobermorite-like or tobermorite because its chemical composition is similar to natural mineral—tobermorite ($5CaO{\cdot}6SiO_2{\cdot}5H_2O$). This phase in general is the entire group of compounds of a hydrated calcium silicate with varying degrees of crystallization and differential alkalinity. The hydration and hardening mechanisms of cement is based on density of C–S–H gel combined with the addition of water and crystallization of ettringite and calcium hydroxide intergrowths forming and filling pores in hardening cement paste. Knowledge of the mineral composition and structure during maturation cement paste allow to predict its technical properties [3].

2 Materials

The hydration process of cement pastes were observed on the basis of mortars samples prepared in accordance with PN-EN 196-1: 1996. To this aim bars about size of 40 × 40 × 160 mm were done, the composition of which was as follows: Portland cement CEM I 42.5 R (450 g), standard sand (1350 g), water (225 g). Thus formed samples were aged in water and subjected to observation of changes in the phase composition with SEM after 2, 28 and 90 days of maturing.

3 Methods

The chemical and mineral composition of products hydration were determined by using a scanning electron microscope SEM-EDS and X-ray method. The morphological forms and the chemical composition of the main mineral components were analysing by means of a scanning electron microscope (SEM) FEI Quanta 250 FEG equipped with a system of chemical composition analysis based on energy dispersive X-ray—EDS of EDAX company [4].

The mineral composition was determined via powder X-ray diffraction (XRD) method using a Philips X'pert APD diffractometer (with a PW 3020 goniometer), Cu lamp, and a graphite monochromator. The analysis was performed within the angle range of 5–65 2θ. Philips X'Pert Highscore software was used to process the diffraction data. The identification of mineral phases was based on the PDF-2 release 2010 database formalized by the ICDD [5].

4 Results and Discussion

The hydration process of cement and development of cement paste structure can be presented in three stages, depicted in Fig. 1 [6].

In the first stage of cement hydration the calcium hydroxide (portlandite) is separated. Its formation is a result of the hydrolysis of tricalcium (C_3S) and dicalcium (C_2S) silicates, what has a place after a few hours of mixing Portland cement with water. The resulting calcium hydroxide at solid phase of paste is present primarily in the form of portlandite. It creates massive, hexagonal crystals about size of 40 microns (Fig. 2), which aggregates taking the form of column (Fig. 3). The morphology of the resulting portlandite crystals is dependent on available free space for crystallization (w/c), the type of admixtures and additives [7]. Portlandite in hydrated paste of Portland cement is up to 25 % by volume of solid phase.

In the second period of hydration of cement paste the first forms of hydrated calcium silicates are created. Their quantitative contents in completely hydrated paste of Portland cement is between 50 and 60 % of the volume of all solid phases. Proposed by Diamond [8] model of the morphological distribution of C–S–H gel in cement paste distinguish form of fiber from size of 2 microns—characteristic for the early stages of hydration (Fig. 4), which goes in the form of a mesh, so-called "honeycomb" (Fig. 5). Subsequent stages of maturing the paste constituted forms becoming more and more massive consisting of a packed and interspersed with each other thin crystals (so called foils), up to the formless and massive gel characteristic for old pastes.

In the third period of paste hydration occurs pore filling of hardening cement paste by short fibers or lamellar phases of hydrated calcium silicates. The duration of this stages is a period of several days to several months, and covers almost complete hydration of the cement. A characteristic feature of this phase is transformation of calcium aluminate trisulfate $3CaO \cdot Al_2O_3 \cdot 3CaSO_4 \cdot 32H_2O$ to calcium aluminate monosulfate $3CaO \cdot Al_2O_3 \cdot CaSO_4 \cdot 12H_2O$. Ettringite crystals and AFt phases typically form an elongated crystals about circular habit similar to the needle

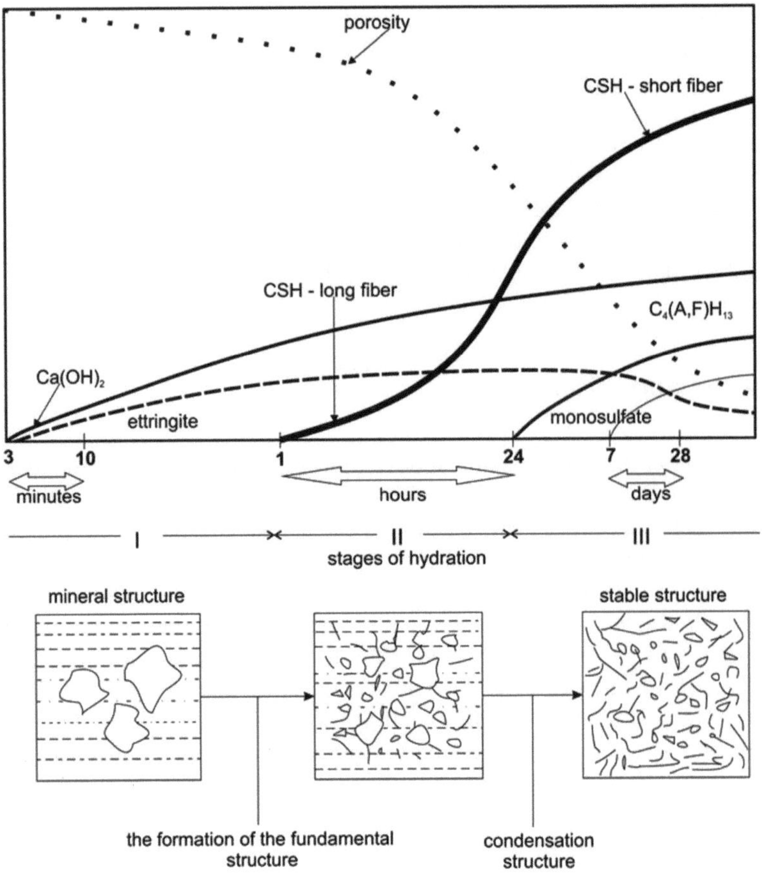

Fig. 1 Formation of the of hydration products and the creation of cement paste structure

Fig. 2 Hexagonal plate of portlandite

Fig. 3 Column aggregate of portlandite

Fig. 4 Needle form of C–S–H and plates of portlandite

Fig. 5 Transformations of portlandite into solid forms, and needle forms of C–S–H into "honeycomb" structure

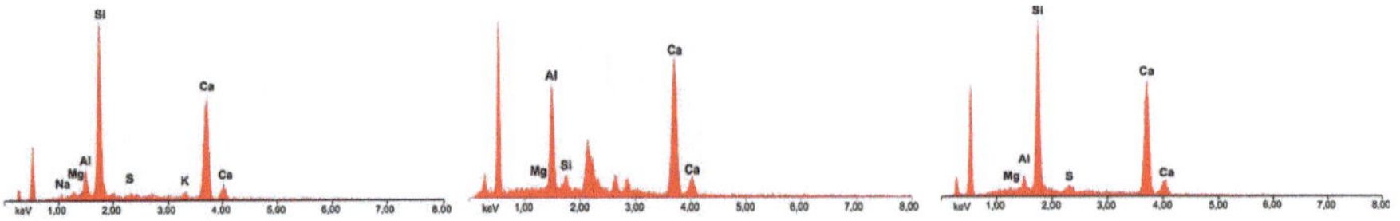

EDS of C-S-H phase EDS of portlandite EDS of C-S-H honeycomb

Fig. 6 Mineral phases occurring in maturing cement paste

Fig. 7 Needle crystals of ettringite

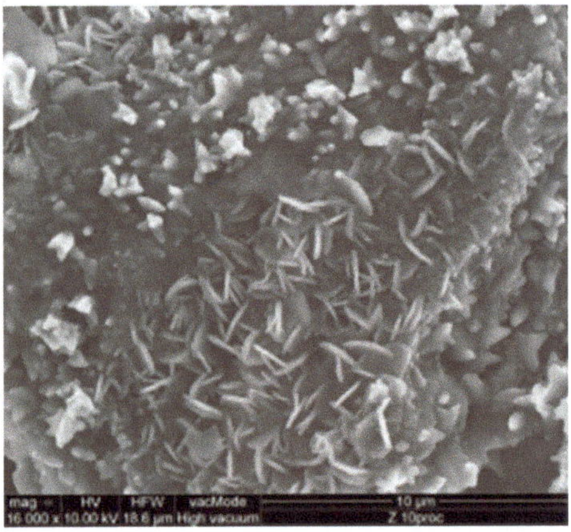

Fig. 8 plate crystals of tobermorite

(Fig. 7), while tobermorite create lamellar aggregates (Fig. 8). The spectra of the chemical composition of the main phase formed during the maturation of the cement paste is shown in Fig. 6.

Fig. 9 Octahedral forms of periclase

Scanning microscope observations also allow to identify the minerals forming the relics of clinker for example periclase that the presence in the paste in the form of large crystals unfavorably affects the stability of the structure of maturing cement pastes (Fig. 9). These types of mineral phases due to their trace amounts in tested samples are not recognizable by diffraction (XRD), or thermal (DTA/TG) methods.

5 Conclusion

The presented results obtained by means of scanning electron microscopy with chemical microanalysis (SEM-EDS) indicate the practical opportunity to observe changes in the microstructure of maturing cement pastes or concrete. Based on changes in the morphology of the main phases arising during the hydration of cement paste it is possible to predict the properties of cement mortars and concretes.

Acknowledgments This research was financed by PL-BY-UA within the Project No. IPBU.01.01.00-06-570/11-00 and Operational Program within the Project No. POKL.04.03.00-00-129/12.

References

1. T.C. Hansen, Physical structure of hardened cement paste. Class. Approach Mater. Struct. **19** (6), 423–436 (1986)
2. M. Arandigoyen, J.I. Alvarez, Pore structure and mechanical properties of cement–lime mortars. Cem. Concr. Res. **37**(5), 767–775 (2007)
3. W. Kurdowski, Chemia cementu i betonu. Wyd. PWN, pp. 1–727 (2010)
4. M. Wdowin, M. Franus, R. Panek, L. Bandura, W. Franus, The conversion technology of fly ash into zeolites. Clean Technol. Environ. Policy **16**(6), 1217–1223 (2014)

5. W. Franus, K. Dudek, Clay minerals and clinoptilolite of Variegated Shales Formation of the Skole Unit. Polish Flysch Carpathians. Geol. Carpath. **50**, 23–24 (1999)
6. R. Locher, W. Richartz, Study of hydration mechanism of cement. Symposium on the chemistry of cement, Moscow, 1974
7. P.K. Mehta, *Concrete, Structure, Properties and Materials* (Prentice-Hall, Englewood Cliffs, 1986)
8. S. Diamond, The microstructures of cement paste in concrete. Proceedings of the 8th International Congress on the Chemistry of Cement, vol. 1 (Rio de Janeiro, 1986), pp. 113–121

Some Properties of 1.3343 Steel Treated by Pulse Plasma Technique

Tuğçe Gökkaya and Yıldız Yaralı Özbek

Abstract This method can be used for the surface modification treatment of materials producing different surface properties. This paper presents the results obtained from an experimental work investigating the changes in some mechanical and microstructural surface properties of a 1.3343 steel (M2 high speed steel). In this work, a battery capacity of 800 μF, and four different sample plasma gun nozzle distances were used. The consumable electrode is tungsten in this work. The samples were subjected to micro-hardness measurements and it was found that the hardness values of modified surfaces are five-seven times higher than that of untreated samples. The modified surface layers were examined using an optical microscope and X-ray analyses were carried out for all samples. It was observed that surface treatment had led to the formation of new phases in surface layers. SEM and EDS analyses were applied to modified surface.

1 Introduction

Irradiation of solids with high-intensity pulses of ion or plasma beams combines the features of three different processes: vacuum evaporation (large fluxes), ion implantation (penetration of ions beneath the surface), and laser melting (energy density sufficient to melt the surface). As a rule, the use of intense energy pulses leads to morphology changes of the treated surfaces [1–4].

Pulse plasma is a surface-hardening process and a well-known emerging technique used to improve the tribological properties, the fatigue endurance and the corrosion resistance of treated work pieces. It has many advantages (short treatment time, low energy use) compared to the traditional surface processes. However, in comparison to the conventional direct current (dc) ion nitriding, some complex

T. Gökkaya · Y.Y. Özbek (✉)
Department of Metallurgy and Materials Engineering, Esentepe Campus, Sakarya University, 54187 Sakarya, Turkey
e-mail: yyarali@sakarya.edu.tr

E.K. Polychroniadis et al. (eds.), *2nd International Multidisciplinary Microscopy and Microanalysis Congress*, Springer Proceedings in Physics 164,
DOI 10.1007/978-3-319-16919-4_15

workpiece shapes with an excellent surface quality can be obtained by the pulse plasma.

When steel is exposed to nitrogen at elevated temperature using different techniques various nitride phases are formed depending on the process conditions. Previous work has shown that when intense pulsed plasma beams were used for surface modification the energy density of the pulse was high enough to melt the near surface region of the substrate and nitrogen and tungsten were introduced into the liquid alloy. The nitrogen expanded austenite is an interstitial solution of nitrogen in fcc iron phase and lattice parameter is expanded in dependence on N and W content. Of great importance is the stability of the phases, predominantly of the expanded austenite phase. At elevated temperature the interstitial atoms can be released from the lattice and diffuse [4, 5]. Depending on the pulse plasma parameters and the base material composition, a compound zone (white layer) and can be formed on the diffusion zone, with thickness in the micron range. In the diffusion zone, nitrogen is held in solid solution within the iron matrix [6].

2 Experimental

The 1.3343 samples were used to modified with different pulse plasma process parameters. The chemical composition of the steel was shown in Table 1. The samples surfaces were ground by from 60 mesh to 600 mesh grid emery papers. The pulse plasma techniques were carried out to all surfaces. The tungsten electrode was selected for consumable electrode. The parameters of pulse plasma treatment were shown in Table 2. The time is short (μs range) in plasma pulses. The energy was high enough to melt the near surface layer. The cooling rate was in the range of 10^7–10^8 K/s. The samples were characterized by the following methods: optic microscope, scanning electron microscopy (SEM), X-ray diffraction analysis (XRD). The microhardness testing was applied from surface to inner.

3 Results and Discussion

The Fig. 1 shows the surface modified by pulse plasma treatment. As a result of surface modification process we obtained thin layers. The three different zone become on modified zone. The thickness of these zone was changed with parameters of process. When increasing the number of pulse, amounts of ionization gases increased on surface. So, the thickness of modified layer was increased. However, the nozzle distance was effected to thickness of modified zone. The optimum thickness of modified layer was obtained in 50 mm distance.

Figure 2 shows SEM micrographs of sample 6 (40 mm distance, 15 pulse). The modified zone is seen easily. The XRD results of untreated surface and treated sample 8 (50 distance-10pulse) is shown in Fig. 3a, b. These figures shows

Table 1 The chemical composition of 1.3343

%	C	Si	Mn	P	S	Cr	Ni	Mo	Al
	0.597	0.301	0.271	0.0122	0.0014	3.88	0.505	4.77	0.0179
Co	Cu	Nb	Ti	V	W	Pb	Sn	B	Fe
0.567	0.136	0.0101	0.0017	1.59	~6.30	0.0017	0.0109	0.0027	81

Table 2 The parameters of pulse plasma treatment

No.	Distance (mm)	Number of pulse	Battery capacity (mf)
1	30	5	800
2	30	10	800
3	30	15	800
4	40	5	800
5	40	10	800
6	40	15	800
7	50	5	800
8	50	10	800
9	50	15	800
10	60	5	800
11	60	10	800
12	60	15	800

significant changes in phase structure as compared with unmodified material. The new phases were occurred after treatment. Released interstitial nitrogen and tungsten atoms become available for the nitrogen and tungsten phases' formation.These phases were important for surface properties such as FeN, Fe_2N, W.

The phase and structure transformations result in the modification of surface characteristics. It was also found that there were some tungsten, nitride and austenite phases on modification layer. The amount of these strong. Figure 4 show us the changing of microhardness values of modified surface. The increase of the pulse numbers or the intensity of energy density absorbed by the surface leads to the growth of the microhardness value. Another reason for the increase in hardness values of the samples was the decreased grain size due to the increased cooling rate within the cycles of pulses. The Guan and friends were synthesized using high ion energies, making use of strain hardening by high defect densities where the defects responsible for the residual stresses also act as obstacles for dislocation movement. The phases precipitated was one of the reasons in leading to increase in hardness values of the treated samples [6, 7]. For example, the surface hardness of modified layer of sample 9 is 5–6 times higher (1475 HV) than that of the untreated samples (250 HV).

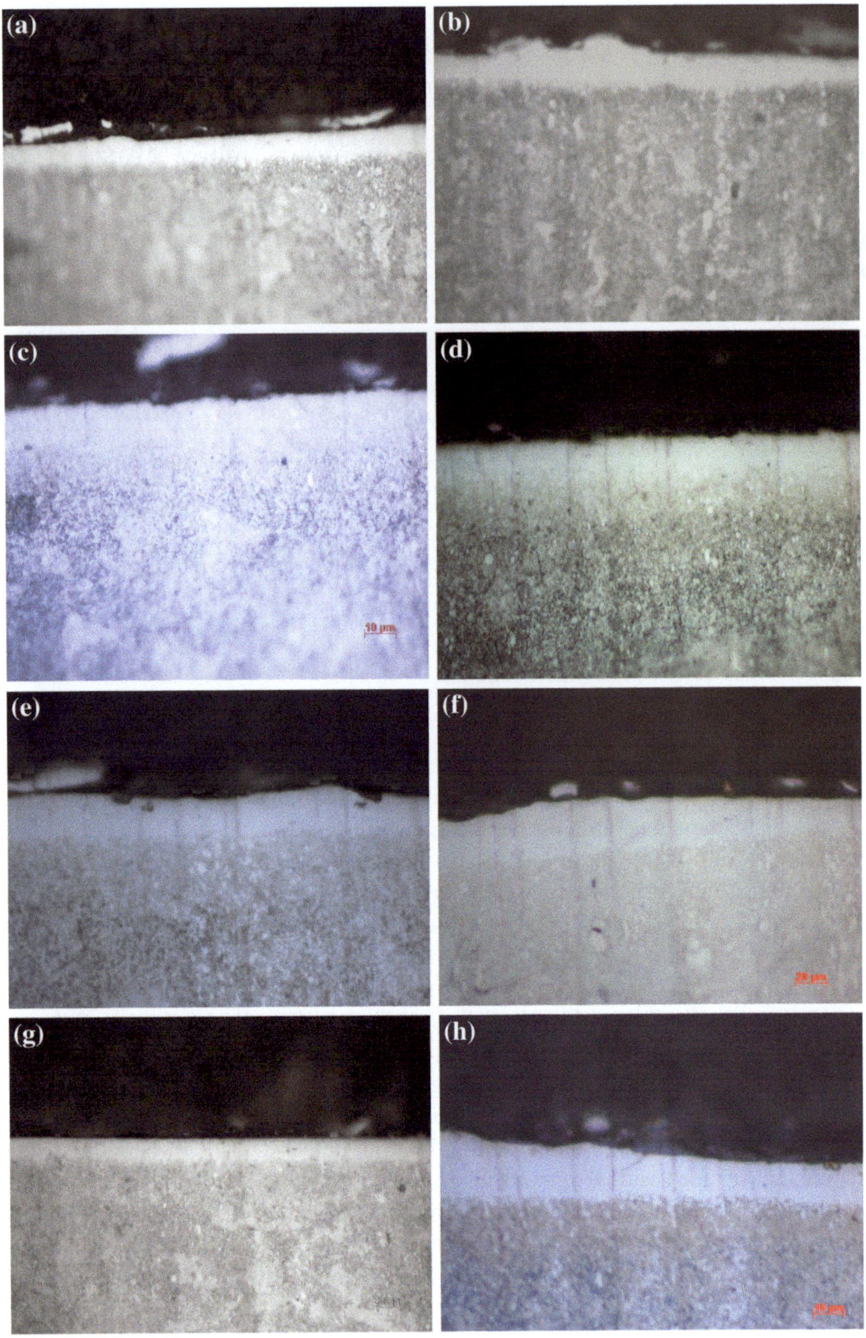

Fig. 1 The optic micrograph of **a** sample 1, **b** sample 2, **c** sample 4, **d** sample 5, **e** sample7, **f** sample 8, **g** sample 10, **h** sample 11

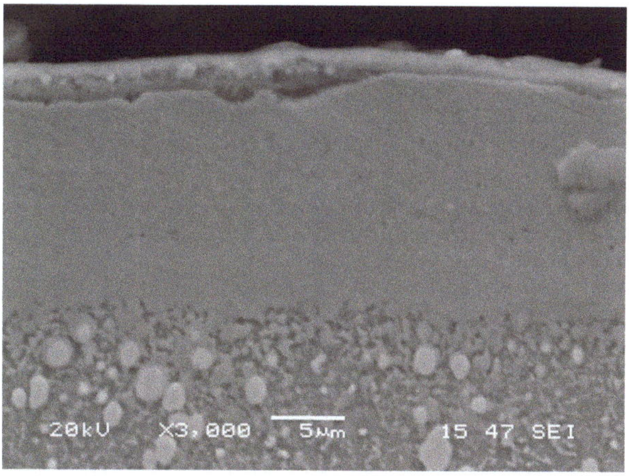

Fig. 2 SEM micrograph of sample 6 (40 mm distance-15 pulse)

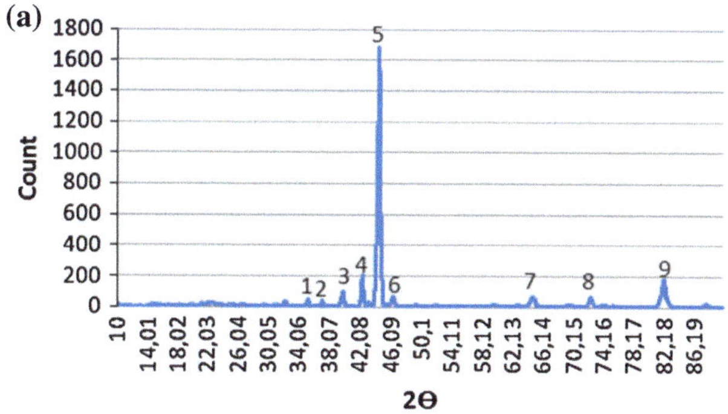

No	Phases
1	Fe_3O_4
2	Fe_3O_4
3	Fe2N
4	W
5	a--Fe,
6	γ-Fe,FeN
7	W
8	γ-Fe,
9	Fe_2N

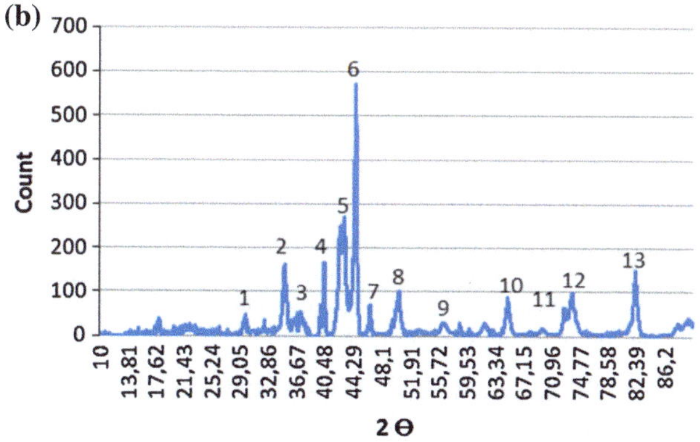

No	Phases
1	Fe_3O_4
2	Fe_3O_4
3	Fe_2N
4	W
5	Fe_2N, γ-Fe, FeN
6	α-Fe
7	γ-Fe,FeN
8	W
9	Fe_2N
10	Fe
11	Fe_3O_4
12	Feγ, FeN
13	Fe, Fe_2N

Fig. 3 XRD results of **a** untreatment sample, **b** sample 8 (50 mm distance, 10 pulse)

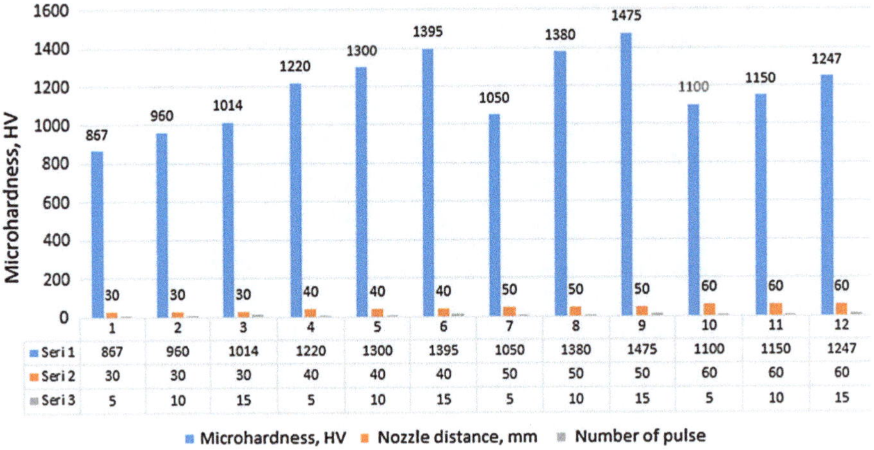

Fig. 4 The microhardness values of surface modified after pulse plasma treatment

4 Conclusion

The pulse plasma treatment leads to the essential modification of structure-phase composition of 1.3343 (M2 steel) and the formation of new phase. The modified zone was formed on surface. The modified layer consists of hard phases after pulse plasma treatment. The concentration of this new phase in modified layer depends on treatment parameters. The observed changes of mechanical properties in steel surface are in good agreement with structure-phase transformations. The mechanical properties are affected to the grain size. The high rate heating and cooling caused thin grain size. The surface hardness values have increased up to 1500 HV by means of this technique in just 1–2 min. These modified steels have very good economical merits.

References

1. J. Piekoszewski, Z. Werner, J. Langner, L. Waliś, Modification of the surface properties of materials by pulsed plasma beams. Surf. Coat. Technol. **106**(2–3), 228–233 (1998)
2. X.P. Zhu, F.G. Zhang, Y. Tang, J.P. Xin, M.K. Lei, Nuclear instruments and methods in physics research section B: beam interactions with materials and atoms. in *Proceedings of the 17th International Conference on Ion Beam Modification of Materials*, vol. 272, 1 Feb 2012, pp. 454–457 (IBMM, 2010)
3. Dynamic response of metals under high-intensity pulsed ion beam irradiation for surface modification. in *The International Symposium on Ion Implantation and Other Applications of Ions and Electrons* (2002)
4. B. Sartowska, J. Piekoszewski, L. Waliś, M. Kopcewicz, Z. Werner, J. Stanisławski, J. Kalinowska, F. Prokert, Phase changes in steels irradiated with intense pulsed plasma beams. Vacuum **70**(2–3), 285–291 (2003)

5. X.Y. Le, W.J. Zhao, S. Yan, B.X. Han, W. Xiang, The thermodynamical process in metal surface due to the irradiation of intense pulsed ion beam. Surf. Coat. Technol. **158–159**, 14–20 (2002)
6. M. Keddam, Surface modification of the pure iron by the pulse plasma nitriding: application of a kinetic model. Mater. Sci. Eng. A **462**(1–2), 169–173 (2007)
7. B. Sartowska, J. Piekoszewski, L. Waliś, J. Senatorski, J. Stanisławski, R. Ratajczak, L. Nowicki, M. Kopcewicz, F. Prokert, M. Barlak, Structural and tribological properties of carbon steels modified by plasma pulses containing inert and active ions. Surf. Coat. Technol. **201**(19–20), 8295–8298 (2007)

Investigation of Microstructure of Ceramics Produced from Gabbro and Zeolite Raw Materials

Nuray Canikoğlu and A. Şükran Demirkıran

Abstract In this study, usage of gabbro and zeolite in the manufacturing of ceramic products were investigated. Three different compositions which consist of zeolite and gabbro as starting materials were prepared to manufacture a ceramic product. Dry mixing process was performed by ball milling for 10 h with alumina balls for all compositions. Compacted samples with 20 mm in diameter were obtained by dry pressing at the pressure of 400 MPa. Then, pressed samples were sintered at 900, 1000, 1100, 1150 and 1200 °C for 1 h in atmospheric conditions. Having completed the sintering process, the microstructural properties of products were examined by optic microscopy and SEM-EDS. The sintering behaviour of the samples produced at different firing temperatures was evaluated by the determination of the water absorption, bulk density and apparent porosity. Also, hardness of the samples was measured.

1 Introduction

In recent years, many of the studies related to stoneware are relevant with the substitution of the traditional raw materials with other low cost minerals and/or wastes. Therefore, rocks of volcanic origin such as nepheline syenite, gabbro, granite, basalt and basaltic tuffs were used in the production of vitrified ceramic tiles fired at temperature range of 1100–1300 °C [1]. The fine-grained gabbro mainly consists of plagioclase and hornblende and contains biotite, apatite, titanite and quartz [2].

N. Canikoğlu (✉) · A.Ş. Demirkıran
Engineering Faculty, Department of Metallurgy and Materials Engineering,
Sakarya University, 54187 Sakarya, Turkey
e-mail: nurayc@sakarya.edu.tr

A.Ş. Demirkıran
e-mail: dkiran@sakarya.edu.tr

E.K. Polychroniadis et al. (eds.), *2nd International Multidisciplinary Microscopy and Microanalysis Congress*, Springer Proceedings in Physics 164,
DOI 10.1007/978-3-319-16919-4_16

Zeolites form a large family of aluminosilicates which have been studied by mineralogists more than 200 years. There are many natural zeolites identified in the world. Clinoptilolite, mordenite, phillipsite, chabazite, stilbite, analcime and laumontite are very common forms whereas offretite, paulingite, barrerite and mazzite are much rarer. In zeolite structure, three relatively independent components are found: the aluminosilicate framework, exchangeable cations, and zeolitic water. The general chemical formula of zeolites is $M_{x/n}[Al_xSi_yO_{2(x+y)}]\cdot pH_2O$ where M is (Na, K, Li) and/or (Ca, Mg, Ba, Sr), n is cation charge; y/x = 1−6, p/x = 1−4 [3]. Zeolites are microporous crystalline solids with well-defined structure that offer large (internal and external) surface areas, consisting of a three-dimensional network of silicon-oxygen SiO_4 and aluminum-oxygen AlO_4 tetrahedra. The high surface area is the basis of high reactivity. Cliniptilolite and heulandite are the most common types of natural zeolite minerals on the earth [4–6]. For their chemical, physical and structural characteristics, zeolites are suitable for a number of applications in various fields such as adsorption of cations, separation, ion exchange, catalysis, building industry, agriculture, soil remediation, and energy [3, 7].

Low-cost natural zeolites are abundantly found as raw materials in many countries; they are very easily crushed and milled; and they facilitate sintering in the production of ceramic bodies [8, 9]. The aim of this study is to produce ceramic products using gabbro and zeolite as starting materials and to examine the properties of obtained products.

2 Experimental Procedure

In present study, three different mixtures which consist of zeolite and gabbro as starting materials were prepared. The chemical compositions of gabbro and zeolite used in the mixtures were given in Table 1. Also, the content of mixtures and marking systems are shown in Table 2.

Firstly, gabbro and zeolite were grinded by dry ring milling and sieved to pass through −75 μm. Then, dry mixing process for homogenization was performed by ball milling for 10 h with alumina balls for all compositions. Disk samples with

Table 1 The chemical compositions of used raw materials

Component (%)	SiO_2	Al_2O_3	TiO_2	Fe_2O_3	CaO	MgO	Na_2O	Others
Gabbro	51.35	14.13	1.96	13.63	7.78	4.70	3.63	2.52
Zeolite	79.28	11.22	0.08	1.20	2.52	1.22	0.15	4.33

Table 2 The codes of samples

Compositions (wt%)	Codes
25 Zeolite–75 Gabbro	25ZG
50 Zeolite–50 Gabbro	50ZG
75 Zeolite–25 Gabbro	75ZG

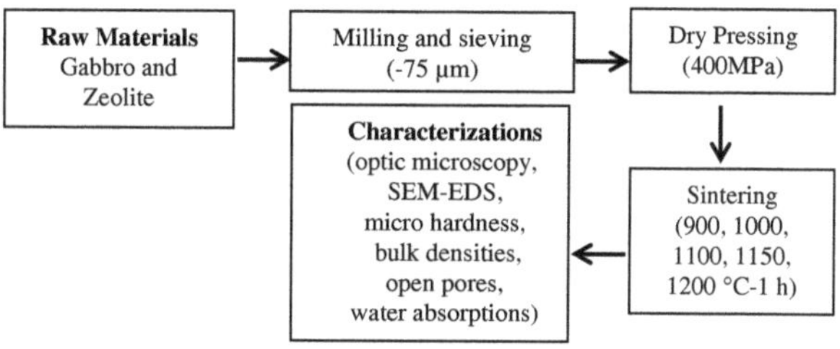

Fig. 1 The flow charts for experimental procedures

20 mm in diameter were prepared by uniaxial dry pressing at the pressure of 400 MPa. After shaping, samples were dried at 110 °C for 24 h in an oven. Dried samples were sintered in an electric furnace with a heating rate of 10 °C/min at 900, 1000, 1100, 1150 and 1200 °C for 1 h in atmospheric conditions. The sintered products were characterized. Experimental procedures are given in Fig. 1.

3 Results and Discussion

The macro photographs of the samples with 25ZG, 50ZG, 75ZG code sintered at 900–1000–1100–1150–1200 °C for 1 h are seen in Fig. 2. As seen in Fig. 2, the sintered samples in low temperatures (900–1000 °C) have light color and are not fully sintered. On the contrary, the shape of the samples which sintered at 1200 °C, distorted because of the sagging and deformation due to the increase in the glassy phase. Therefore, characterisation of samples sintered at 900–1000 and 1200 °C was not attempted. Thus, the suitable sintering temperatures of the samples were determined to be 1100 and 1150 °C.

Hardness of the samples sintered at 1100–1150 °C was measured using a Vickers microhardness tester. The microhardness variation of the samples as a function of sintering temperature is given in Fig. 3a. As seen in Fig. 3a, the hardness values increased with increasing the sintering temperature. The maximum hardness value was measured in 25ZG coded sample sintered at 1150 °C as 823 HV. Additionally, the decrease of gabbro amount decreased the hardness.

The bulk densities of sintered samples were measure using Archimedes method. The bulk density variation of samples depending on the increase of zeolite quantity is illustrated in Fig. 3b. As seen from this figure, the bulk density values increased with increasing both the addition of gabbro and the sintering temperature. The maximum bulk density value was found in 25ZG coded sample sintered at 1150 °C as 2.33 g/cm^3.

As is known, water absorption is an important indication for densification. The apparent porosity and water absorption variations of all compositions were shown in Fig. 4a, b. The both of values decreased with increasing sintering temperature.

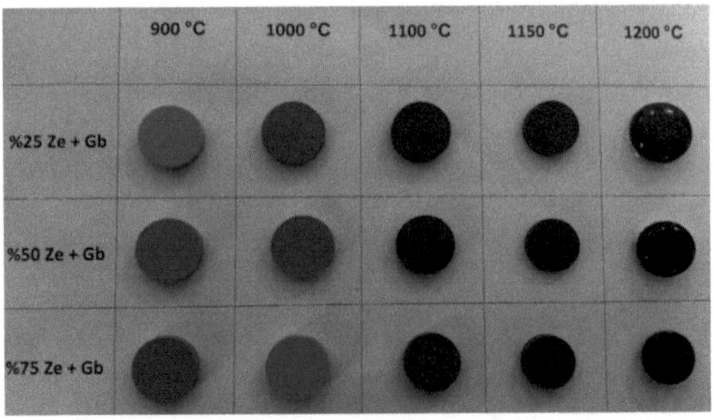

Fig. 2 The macro photographs of sintered samples

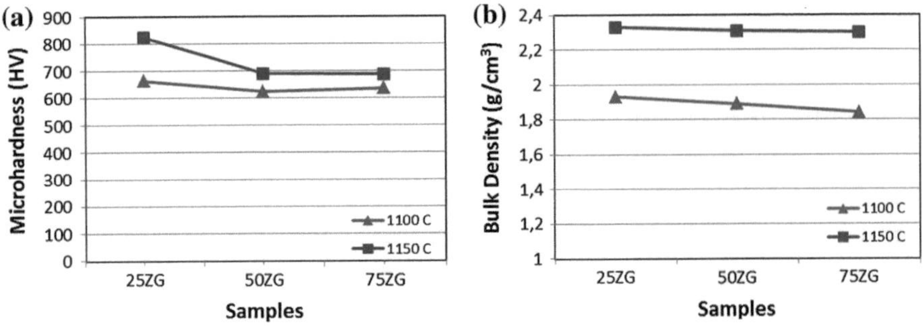

Fig. 3 Microhardness variation (**a**), the bulk density values (**b**) of samples sintered at 1100 and 1150 °C for 1 h

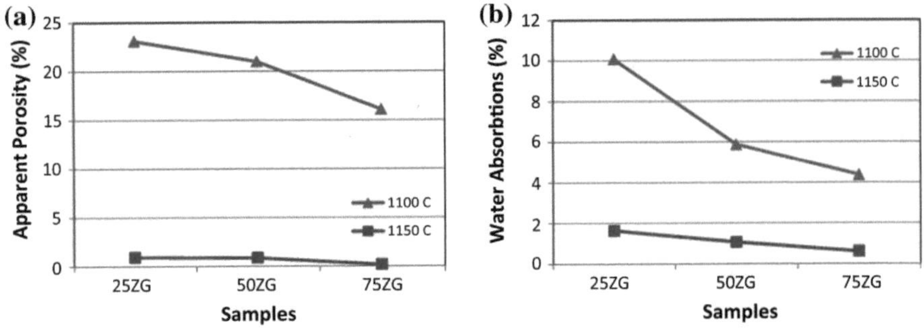

Fig. 4 Apparent porosity (**a**), water absorption variation (**b**) of samples sintered at 1100 and 1150 °C for 1 h

This demonstrated that the ideal sintering temperature of the samples is 1150 °C. Additionally, the increase of gabbro amount increased the apparent porosity and the water absorption.

The optical micrographs of samples coded 25ZG, 50ZG and 75ZG which were sintered at 1100 and 1150 °C for 1 h, are shown in Fig. 5. Compared to micrographs, the amount of porosity is more in the sintered samples at 1100 °C. The densification increases with the increase in sintering temperature. This result is confirmed with the bulk density graphic given in Fig. 3b and it is shown that sintering process at 1150 °C is more successful. As the results of all the analysis, the 25ZG sample sintered at 1150 °C has all desired properties.

In the SEM examination of the 25ZG sample sintered at 1150 °C given in Fig. 6 were seen very small cracks. It is thought that, these cracks are due to from thermal expansion mismatch between different phases. These results are similar to the literature [8]. Also, it is shown that a phase precipitated on the matrix structure.

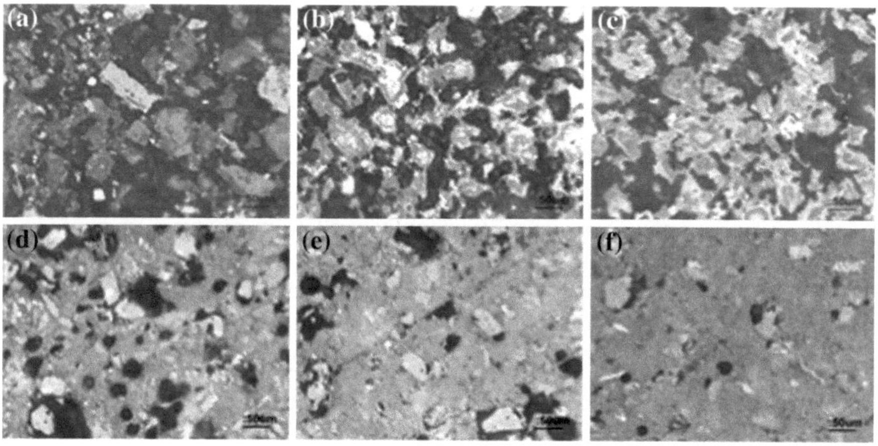

Fig. 5 The optic photographs of samples sintered at 1100 °C for 1 h **a** 25ZG, **b** 50ZG, **c** 75ZG and the optic photographs of samples sintered at 1150 °C for 1 h **d** 25ZG, **e** 50ZG, **f** 75ZG

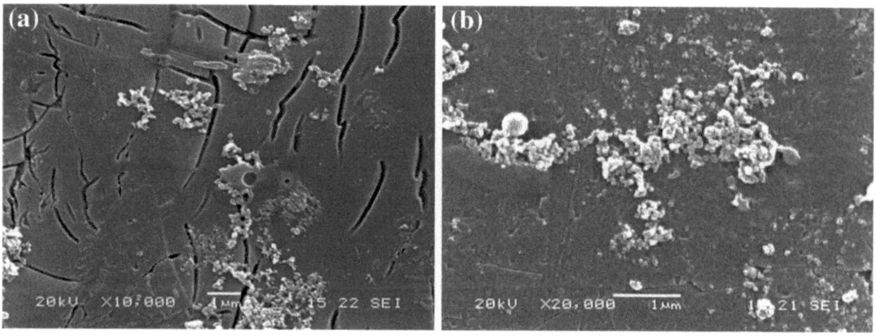

Fig. 6 The SEM micrographs (**a**) and (**b**) of the 25ZG sample sintered at 1150 °C for 1 h

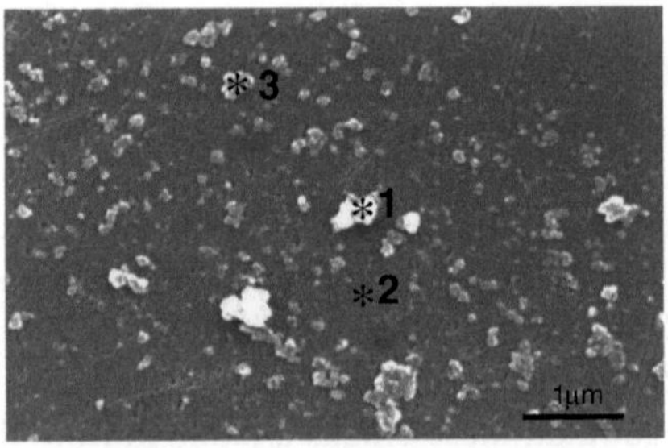

Elements	1	2	3
O	38.641	35.834	33.953
Na	4.252	4.455	4.387
Al	14.318	14.216	14.836
Si	37.003	39.364	39.384
K	2.352	2.920	3.651
Ca	3.435	3.211	3.790

Fig. 7 The EDS analysis of the 25ZG sample sintered at 1150 °C for 1 h

The EDS analysis was applied for tree different regions and Si, O, Al, K, Ca, Na elements were detected. These elements belong to chemical compositions given in Table 1. The EDS analyses were presented in Fig. 7.

4 Conclusions

In this study, ceramic products were manufactured from mixtures prepared with gabbro and zeolite economically. Prepared mixes were sintered at 900, 1000, 1100, 1150 and 1200 °C for 1 h and later, microstructure and some properties of sintered samples were examined. The following conclusions can be drawn from this study.

- Hardness, bulk density, water absorption and apparent porosity measurements of the sintered samples revealed that 900 and 1000 °C are not sufficient sintering temperatures. For a sintering temperature of 1200 °C, deformation and sagging of the samples was observed. The ideal sintering temperature for samples with zeolite and gabbro was determined to be 1150 °C.
- The hardness values of the 25ZG, 50ZG and 75ZG samples produced at 1150 °C were 823, 690 and 687 HV, respectively. Also, the bulk density values of the 25ZG, 50ZG and 75ZG samples produced at 1150 °C were 2.33, 2.31 and 2.30 g/cm^3, respectively.
- While the hardness and bulk density of the samples sintered at 1150 °C increased with increasing gabbro, the apparent porosity and water absorption decreased with increasing gabbro addition.

References

1. S.M. Naga, F. Bondioli, M.M.S. Wahsh, M. El-Omla, Utilization of granodiorite in the production of porcelain stoneware tiles. Ceram. Int. **38**, 6267–6272 (2012)
2. K. Horie, Y. Tsutsumi, M. Cho, Y. Morishita, H. Hidaka, Crystallization of REE minerals and redistribution of U, Th, and REE at contact boundary between granite and gabbro during hydrothermal alteration. Phys. Chem. Earth **35**, 284–291 (2010)

3. S. Wang, Y. Peng, Natural zeolites as effective adsorbents in water and wastewater treatment. Chem. Eng. J. **156**, 11–24 (2010)
4. Y. Kocak, E. Tascı, U. Kaya, The effect of using natural zeolite on the properties and hydration characteristics of blended cements. Constr. Build. Mater. **47**, 720–727 (2013)
5. T. Perraki, A. Orfanoudaki, Mineralogical study of zeolites from Pentalofos area, Thrace, Greece. Appl. Clay Sci. **25**, 9–16 (2004)
6. K. Elaiopoulos, T. Perraki, E. Grigoropoulou, Mineralogical study and porosimetry measurements of zeolites from Scaloma area, Thrace, Greece. Microporous Mesoporous Mater. **112**, 441–449 (2008)
7. P. Castaldi, L. Santona, S. Enzo, P. Melis, Sorption processes and XRD analysis of a natural zeolite exchanged with Pb^{2+}, Cd^{2+} and Zn^{2+} cations. J. Hazard. Mater. **156**, 428–434 (2008)
8. A.Ş. Demirkiran, R. Artir, E. Avci, Effect of natural zeolite addition on sintering kinetics of porcelain bodies. J. Mater. Process. Technol. **203**, 465–470 (2008)
9. R. Gennaro, P. Cappelletti, G. Cerri, M. Gennaro, M. Dondi, G. Guarini, A. Langella, D. Naimo, Influence of zeolites on the sintering and technological properties of porcelain stoneware tiles. J. Eur. Ceram. Soc. **23**, 2237–2245 (2003)

The Effect of MgO and MgO–Al_2O_3 on Zirconia Produced by Precipitation Method

M. Ipek

Abstract The aim of this study is to produce partially stabilized zirconia with MgO and to investigate effect of Al_2O_3 on partially stabilized zirconia. Zirconia nanopowders are synthesized by the precipitation method from zirconium sulfate precursors. Precipitates are dried at 80 °C for 72 h and 2.77 in weight percentage MgO powders were added to dried powders. Then this powders mixture is calcinated at 1300 °C for 1 h. Furthermore, wt%10 Al_2O_3 is added to calcinated powders and two different powders mixtures are produced. Each powder mixture are pressed and sintered at 1650 °C for 1 and 2 h in an electrical resistance furnace. MgO–ZrO_2 samples have cubic, tetragonal and monoclinic zirconia phases while alumina added MgO–ZrO_2–Al_2O_3 samples have dominantly monoclinic zirconia and spinel phases. In MgO–ZrO_2–Al_2O_3 samples, MgO dissolves in Al_2O_3 and therefore MgO in zirconia matrix is not enough for the stabilization. Microhardness and fracture toughness of MgO–ZrO_2 samples are higher than that of MgO–ZrO_2–Al_2O_3 samples.

1 Introduction

Ceramics from partially stabilized zirconia (PSZ) are undoubtedly one of the greatest scientific discoveries in the field of materials science in the second half of the twentieth century. Soon after the publication of, which described the creation of a material called ceramic steel, hundreds of works were devoted to the phenomenon of transformation toughening, which determines the high level of mechanical properties of PSZ ceramics [1].

Pure zirconia having polymorphic structure is monoclinic (*m*) at room temperature and atmospheric pressure with increasing temperature the material transforms

M. Ipek (✉)
Engineering Faculty, Department of Metallurgy and Materials Engineering, Esentepe Campus, Sakarya University, 54187 Sakarya, Turkey
e-mail: mipek@sakarya.edu.tr

E.K. Polychroniadis et al. (eds.), *2nd International Multidisciplinary Microscopy and Microanalysis Congress*, Springer Proceedings in Physics 164,
DOI 10.1007/978-3-319-16919-4_17

to tetragonal (*t*), approximately at 1170 °C and then to a cubic (*c*) about 2370 °C with melting at 2716 °C. Volume changes on cooling associated with these transformations are substantial enough to make the pure material unsuitable for applications requiring an intact solid structure: $c \rightarrow t$ approximately 2.31 %; $t \rightarrow m$ approximately 4.5 %. Sintered structures transforming from *t* to *m* on cooling from sintering temperatures undergo spallation with portions crumbling into multi-grained powders [2]. The phase changes may be suppressed by stabilization of the high temperature forms by alloying with other oxides. Certain cubic oxides such as CaO, MgO, Y_2O_3, CeO_2 and other rare-earth oxides stabilize the high temperature cubic phase to room temperature and at the same time promote a decrease in the transformation temperatures. However, if insufficient stabilizing oxide is added than the amount required for full stabilization, a partially stabilized zirconia (PSZ) is formed and two or three phases (cubic, tetragonal and monoclinic) microstructure results. The good mechanical properties of PSZ ceramics are primarily as a result of stress induced martensitic phase transformation of metastable tetragonal phase to the stable monoclinic phase, hence the name transformation toughening [2–8].

The aim of this study is to fabricate a ceramic based on MgO partially stabilized zirconia (Mg-PSZ) from precipitated zirconia and MgO powders and to investigate effect of Al_2O_3 on Mg-PSZ.

2 Experimental Details

Zirconia powders are synthesized by precipitation method from $Zr(SO_4)_2$ salt. Zr $(SO_4)_2$ is dissolved in water and NH_4OH is added into solution for precipitation. Precipitate is dried at 80 °C for 72 h and 2.77 in weight percentage MgO powders are added to dried powders and mixed by conventionally ball milling for 2 h. Then, the powder mixture (Mg-PSZ) is calcinated at 1300 °C for 1 h, thus the first powder mixture is prepared. In order to obtain the second powder mixture (Mg-PSZ+A), wt% $10Al_2O_3$ powders are added to calcinated MgO+ZrO_2 powders. Each powder mixture is consolidated by uniaxial pressing at 300 MPa without any binder. The compacts are finally sintered in an open atmospheric electric resistance furnace at 1600 °C for 1 and 2 h with heating and cooling rates of 5 °C/min. Phase analysis of sintered samples are performed XRD instrument by using Cu Kα radiation. The morphology and the microstructure of the products are examined by means of scanning electron microscopy (SEM), energy dispersive spectroscopy (EDS). The microhardness and fracture toughness (K_{Ic}) of the polished specimens is determined applying a Vickers indenter with a load of 1 kg. K_{Ic} is calculated using the (1) [9, 10].

$$K_{Ic} = 0.016\left[\frac{E}{H}\right]^{0.5}\frac{P}{c^{3/2}} \tag{1}$$

where E is Young's modulus, H is hardness and c is the indentation half crack length.

3 Results and Discussion

XRD patterns of Mg-PSZ and Mg-PSZ+A sintered samples at 1600 °C for 1 and 2 h are given in Fig. 1. Mg-PSZ samples have mixture of cubic, tetragonal and monoclinic zirconia phases while alumina added Mg-PSZ-A samples have dominantly monoclinic zirconia and spinel phases ($MgAl_2O_4$). The stabilization effect of magnesia is not observed in Mg-PSZ-A sample and it forms spinel phase with the alumina. The volume fraction of the monoclinic phase in Mg-PSZ is calculated by the following equation [11, 12]

$$V_m = \frac{I_{(\bar{1}11)m} + I_{(111)m}}{I_{(\bar{1}11)m} + I_{(111)m} + I_{(111)c,t}} \tag{2}$$

where V_m is the volume fraction of the monoclinic phase and I is the intensity of specific XRD peaks. The increasing sintering time from 1 to 2 h causes a decrease in the monoclinic phase from 9.83 to 6.62 %. The reason for this change is increase the amount of magnesium dissolved in zirconia with increasing sintering time

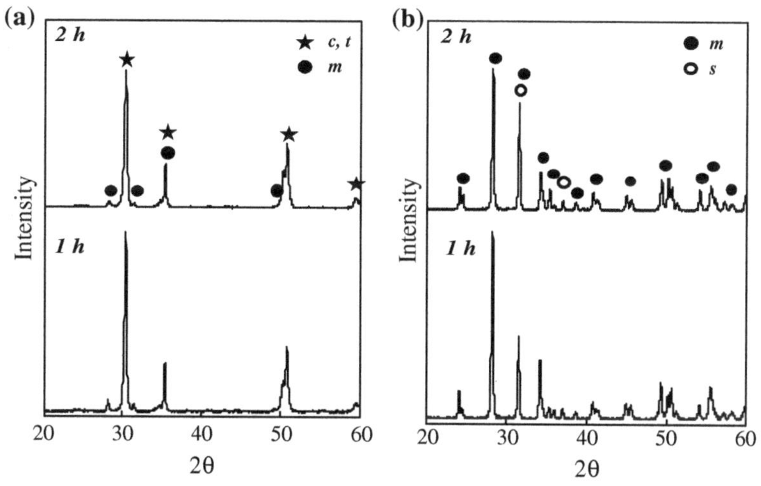

Fig. 1 XRD patterns of sintered samples, **a** Mg-PSZ and **b** Mg-PSZ+A [*c* cubic *t* tetragonal and *m* monoclinic zirconia, *s* spinel ($MgAl_2O_4$)]

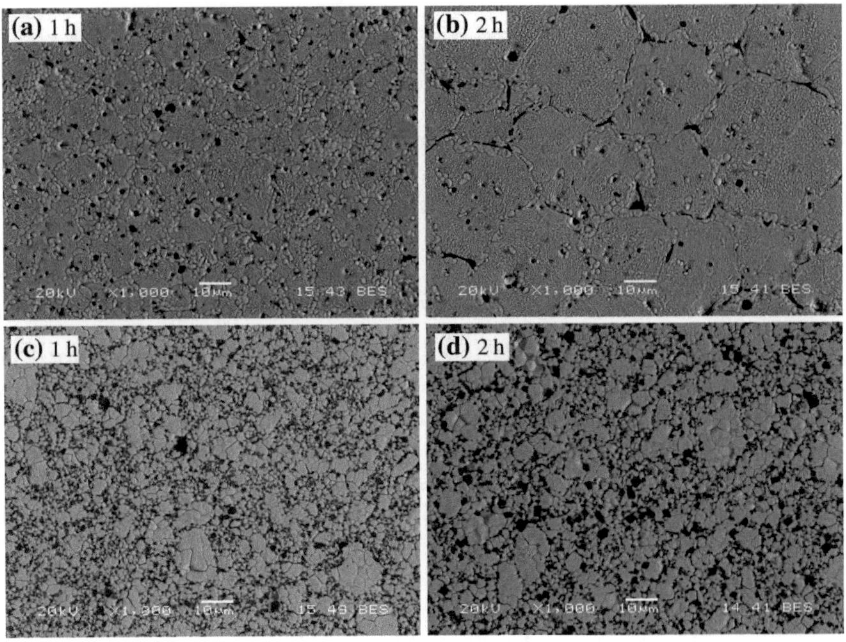

Fig. 2 SEM micrographs of samples sintered at 1600 °C for (**a**) and (**b**) Mg-PSZ, (**c**) and (**d**) Mg-PSZ-A

which leads to increase the volume fraction of the cubic phase or the tetragonal phase [12].

SEM micrographs of samples sintered at 1600 °C for 1 and 2 h, which are polished and thermally etched at 1500 °C for 1 h, are illustrated in Fig. 2. In micrograph of Mg-PSZ samples, light gray and dark gray areas represent zirconia and magnesium rich phases, respectively. Mg-PSZ samples have heterogeneous grain size distribution in such a way that small grains, ~1 µm in size and coarse grains, >10 µm in size. The amount and size of coarse grains increase with increasing sintering time. Coarse grains are cubic (stabilized) zirconia having tetragonal zirconia precipitates which contain 2.77 wt% of Mg (Fig. 3a, Marks 2, 3, 4) and small grains represent the pure zirconia (monoclinic) without Mg stabilizer (Fig. 3a Marks 1, 5, 6). Coarse grains in Mg-PSZ samples include magnesium together with zirconium; it is possible to claim that the dissolution of magnesium in zirconia promotes grain growth of Mg-PSZ [13]. In Mg-PSZ+A samples, light gray areas are zirconium and dark gray areas are aluminum rich grains. Magnesium is dissolved primarily in alumina grains as it can be seen in Fig. 4, Marks 2, 3, 4, 5, 6. This result shows that alumina behaves as a destabilizer in Mg-PSZ [12].

Table 1 shows the hardness and the fracture toughness values of Mg-PSZ and Mg-PSZ-A samples sintered at 1600 °C for 1 and 2 h. Microhardness values of Mg-PSZ samples (approximately 1200 HV) are higher than that of Mg-PSZ-A

(a)

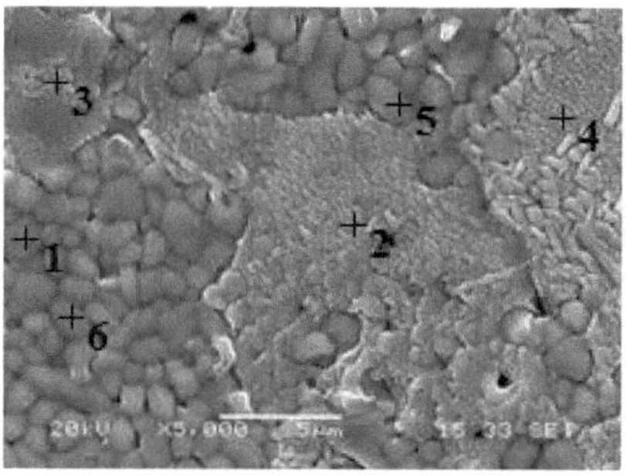

Mark	wt.%		
	O	Mg	Zr
1	24.343	0.207	75.451
2	25.004	2.020	72.976
3	19.020	2.809	78.172
4	19.089	2.770	78.140
5	19.964	0.502	79.535
6	18.714	0.212	81.074

(b)

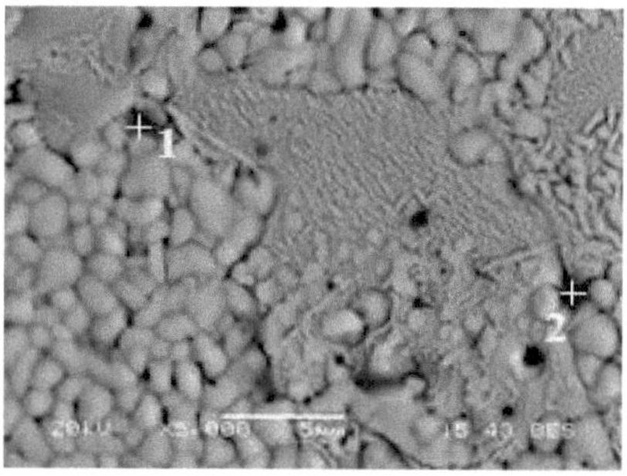

Mark	wt.%		
	O	Mg	Zr
1	27.777	21.097	51.126
2	30.748	18.008	51.244

Fig. 3 SEM micrographs and EDS point analysis of Mg-PSZ sample sintered for 1 h

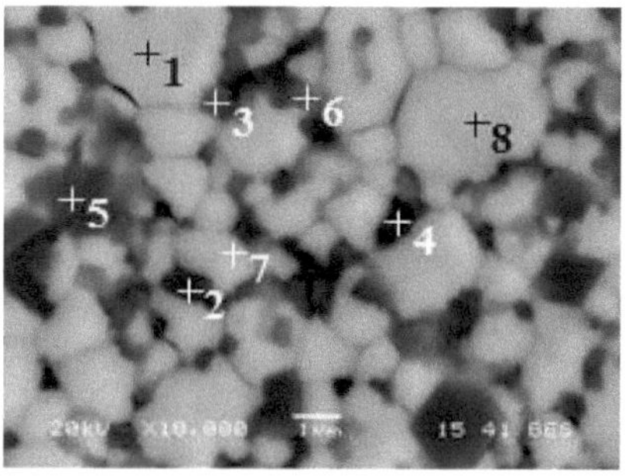

Mark	wt.%			
	O	Mg	Al	Zr
1	20.945	0.326	-	78.729
2	33.456	6.615	17.961	41.969
3	33.171	4.921	11.922	49.987
4	32.578	8.032	22.337	37.052
5	27.879	4.892	13.130	54.099
6	33.663	4.582	12.499	49.256
7	18.876	0.603	1.036	79.485
8	21.868	0.186	-	77.946

Fig. 4 SEM micrographs and EDS point analysis of Mg-PSZ-A sample sintered for 1 h

Table 1 The hardness and the fracture toughness values of sintered samples

	Mg-PSZ		Mg-PSZ-A	
	1 h	2 h	1 h	2 h
Hardness (HV)	1218	1270	570	640
K_{IC}, MPa $m^{1/2}$	3.8	5.5	–	–

samples (approximately 600 HV), also fracture toughness values cannot be determined for Mg-PSZ-A samples because of not having stabilized MgO and therefore they have very brittle structure with monoclinic zirconia phase as determined in XRD and SEM-EDS analysis. Fracture toughness value of Mg-PSZ sample sintered for 2 h is 5.5 MPa $m^{1/2}$ which has higher than sample sintered for 1 h. It is observed that increasing the quantity of cubic and tetragonal phase with increasing sintering time increase the fracture toughness of Mg-PSZ sample sintered for 2 h.

4 Conclusion

The results obtained from this study are listed below.

1. Mg-PSZ samples have mixture of cubic, tetragonal and monoclinic zirconia phases while alumina added Mg-PSZ-A samples have dominantly monoclinic zirconia and spinel phases ($MgAl_2O_4$).
2. The increasing sintering time from 1 to 2 h causes a decrease in the monoclinic phase from 9.83 to 6.62 %.
3. Microhardness values of Mg-PSZ samples (approximately 1200 HV) are higher than that of Mg-PSZ-A samples (approximately 600 HV).
4. Fracture toughness value of Mg-PSZ sample sintered for 2 h is 5.5 MPa $m^{1/2}$ which has higher than sample sintered for 1 h. It is observed that increasing the quantity of cubic and tetragonal phase with increasing sintering time increase the fracture toughness of Mg-PSZ sample sintered for 2 h.

Acknowledgments The author thanks expert Fuat Kayis for performing SEM-XRD studies and technician Ersan Demir for experimental assistance at Sakarya University. Also, special appreciations are extended to Prof. Dr. Cuma Bindal and Prof. Dr. Sakin Zeytin for their notable support.

References

1. G.Y. Akimov, On structural features responsible for the mechanical properties of ceramics from zirconia partially stabilized with Y^{3+} and Mg^{2+} ions. Refract. Ind. Ceram **41**, 427–429 (2000)
2. J.R. Kelly, I. Denry, Stabilized zirconia as a structural ceramic: an overview. Dent. Mater. **24**, 289–298 (2008)
3. R.H.J. Hannink, Growth morphology of the tetragonal phase in partially stabilized zirconia. J. Mater. Sci. **13**, 2487–2496 (1978)
4. M.H. Bocanegra-Bernal, S.D. La Torre, Review phase transitions in zirconium dioxide and related materials for high performance engineering ceramics. J. Mater. Sci. **37**, 4947–4971 (2002)
5. E.A. Korableva, V.S. Yakushkina, O.S. Grishin, V.V. Vikulin, O.P. Dyachenkol, A structural study of Y_2O_3-partially stabilized zirconia ceramics. Refract. Ind. Ceram **46**, 21–23 (2005)

6. I. Denry, J.R. Kelly, State of the art of zirconia for dental applications. Dent. Mater. **24**, 299–307 (2008)
7. R.K. Govila, Strength characterization of MgO-partially stabilized zirconia. J Mater. Sci. **26**, 1545–1555 (1991)
8. M. Ipek, Characterization of CaO–ZrO_2 and CaO–ZrO_2–Al_2O_3. Proceedings of the InterM, 99–104 (2013)
9. S.M. Smith, O.R. Scattergood, Crack shape effects for indentation fracture toughness measurements. J. Am. Ceram. Soc. **75**, 305–315 (1992)
10. D. Sarkar, S. Adak, N.K. Mitra, Preparation and characterization of an Al_2O_3–ZrO_2 nanocomposite. Part I. powder synth. transform. behav. during fract., compos., part A. **38**, 124–131 (2007)
11. P.G. Rao, M. Iwasa, T. Tanaka, I. Kondoh, Preparation and mechanical properties of Al_2O_3–15 wt%ZrO_2 composites. Scr. Mater. **48**, 437–441 (2003)
12. S. Yoon, C.J. Van Tyne, H. Lee, Effect of alumina addition on the microstructure and grain boundary resistance of magnesia partially-stabilized zirconia. Curr. Appl. Phys. **14**, 922–927 (2014)
13. S. Yoon, T. Noh, W. Kim, J. Choi, H. Lee, Structural parameters and oxygen ion conductivity of Y2O3–ZrO2 and MgO–ZrO_2 at high temperature. Ceram. Int. **39**, 9247–9251 (2013)

Microstructural Investigation of IF Steels Joined by Metal Inert Gas Brazing

Mehmet Ekici, Faruk Varol, Uğur Özsaraç and Salim Aslanlar

Abstract In this study, interstitial-free (IF) Erdemir 7116 and galvanized Erdemir 1315 steels having 1.0 mm thickness were joined by metal inert gas (MIG)-brazing method. CuAl8 wire was used as filler material. The specimens were cut in 200 mm × 200 mm dimensions and joined by using 40, 45, 50, 55, 60, 65 and 70 A currents both in butt and overlap joint forms. Having prepared the joints, all specimens were exposed to tensile test according to TS 138 EN 10002-1 standard. Micro and macro-structure investigation of samples were performed by optical microscope and scanning electron microscope (SEM). The matrix structure and the other phases occurred during brazing process were characterized and the effects of them on mechanical properties of joints were discussed.

1 Introduction

IF (Interstitial-Free) steels are produced with very low amounts of interstitial elements (primarily carbon and nitrogen) with small amounts of titanium or niobium added to tie up the remaining interstitial atoms. Without free interstitial elements, these steels are very ductile and soft, will not age or bake harden, and will not form

M. Ekici (✉)
Vocational School of Yalova, Yalova University, Yalova, Turkey
e-mail: mekici@yalova.edu.tr

F. Varol
Vocational School of Karasu, Sakarya University, Sakarya, Turkey
e-mail: fvarol@sakarya.edu.tr

U. Özsaraç · S. Aslanlar
Technology Faculty, Department of Metallurgical and Materials Engineering, Sakarya University, Sakarya, Turkey
e-mail: ozsarac@sakarya.edu.tr

S. Aslanlar
e-mail: aslanlar@sakarya.edu.tr

E.K. Polychroniadis et al. (eds.), *2nd International Multidisciplinary Microscopy and Microanalysis Congress*, Springer Proceedings in Physics 164,
DOI 10.1007/978-3-319-16919-4_18

strain (Lüder's) lines during forming due to the absence of yield point elongation. Today, automotive industry extensively uses room-temperature sheet metal forming processes for the fabrication of car body parts often in complex geometries. Interstitial-free steel (IF-steel) is one of the materials that meet requirements for these forming operations because of its excellent formability and high planar isotropy [1, 2]. IF steels have a single-phase ferritic microstructure with decreased amount of interstitial carbon and nitrogen in the ferrite which is achieved by micro-alloying elements of Ti and/or Nb and formation of very small volume fractions of carbides and nitrides [1]. However, this microstructure brings about a low strength because of the decrease in the solid solution hardening effect of interstitial atoms [1]. Low strength in automotive body panel materials gives rise to the need for thicker sheets in order to maintain desired crash safety and panel rigidity, which cause heavier vehicles, and consequently higher fuel consumption. In addition, low strength of IF-steel limits its broader applications in conditions where high strength is needed in addition to high formability. Enhancing the strength with adequate ductility may increase the potential applications of IF-steel sheets in new applications like those in aviation or defense industries.

There are many studies in literature dealing with the welding of interstitial-free (IF) steels. For example, Saray et al. investigated the mechanical properties of equal-channel angular sheet extruded interstitial-free (IF) steel [1]. Kundu et al. joined the IF steel with commercially pure aluminum by friction-stir welding process and calculated the tensile strength of joints [2]. Khosravi et al. joined galvanized and electro-galvanized interstitial free steel sheets by resistance seam welding in order to see the weldability behaviour of them [3]. Not only resistance seam welding, but also resistance spot welding is an important method in joining of IF steels. Hayat [4] also gained dissimilar joints by joining BH180-AISI 304L Steels and BH180-IF 7123 Steels by using resistance spot welding method. Similarly, Mukhopadhyaya et al. determined the strength values of spot welded IF steel sheets [5]. The results of their study unambiguously infer that the strength values of spot-welds remain same in a specific mode of loading, while the load-bearing capacity increases with increasing nugget size. In addition, Panda et al. characterized the tensile properties of stretchily formed tailor welded IF steels [6]. The durability and structural performance of any auto-body structure is largely dependent on the quality and the design of the component joints. For maximum performance the joint should be designed with parallel consideration given to the joining process, taking into account accessibility of the parts to be joined in the production environment. There are three fundamental options for joining auto-body materials and components: 1-welding, 2-adhesive bonding, and 3-mechanical fastening. By far the most common materials joining process in the automotive industry is spot welding, other joining processes like gas metal arc brazing are being used today. There has been an increasing interest in the use of arc brazing especially for joining of zinc coated steels with low thickness up to about 3 mm. Owing to the high heat input of arc welding there are frequently problems with spatter and porosity connected to zinc melting and evaporation. Zinc has a melting temperature of about 420 °C and vaporises at about 900 °C. This is much lower than the melting point of steel at about 1400 °C. There is also high deformation of the parts during and after welding. One possible way of solving these problems is to use arc brazing that uses lower temperatures for joining. The method is very similar to arc welding with high productivity and the possibility to robotise. The same type of welding equipment can also be used. In arc brazing, the brazing material is melted in an electric arc that also heats the base material. Both MIG, TIG or plasma processes can be used. Robotised MIG brazing is efficient with high speed and good quality. Since the shielding gases can be either inert or with some additions of active gases it is more correct to use the name MIG/MAG brazing. There are less study about MIG-brazing of IF steels in literature. Qin et al. joined aluminium alloy and galvanized steel by laser assisted GMA-brazing [7]. Therefore this study was performed in order to investigate the MIG-brazing ability of IF steels. This present work reports the results of gas metal arc brazing process made by SG-CuSi3 wire used as filler metals in joining of interstitial-free steel sheets and focuses on the microstructure characterization of brazed joints. Microstructural characteristics of the joint region were examined in detail.

2 Experimental Work

2.1 Materials

The interstitial-free (IF) Erdemir 7116 and galvanized Erdemir 1315 steel plates having 1.0 mm thickness were joined by copper-based CuAl8 wire in gas metal arc brazing technique. Specimens were prepared in two different joining forms as butt joint and overlap joint. Both types of brazing operations were done with seven different weld currents as 40, 45, 50, 55, 60, 65, and 70 A. The chemical composition of steel sheets used in brazing operations was given in Tables 1 and 2, the chemical composition of filler material used during brazing operations was given in Table 3.

2.2 MIG-Brazing Operations

The gas metal arc brazing (MIG-brazing) operations were performed in a welding machine which was specially designed for brazing purposes in D.C. (+) position. The shielding gas was chosen as Argon inert gas and 100 % Argon tubes were used during metal inert gas (MIG-brazing) operations. In literature, it was recommended that gas mixtures of CO_2, NO and O_2 with varying amounts can be added into Argon in order to obtain arc stability [8] but we preferred pure Argon to check the joinability of thin IF sheets for automotive applications. Our specially designed and manufactured MIG-brazing machine ensures controlled transfer of the molten metal through the electric arc directed to high quality requests and joints of steel sheets

Table 1 Spectrometric analysis and the mechanical properties of galvanizeless 7116 IF steel sheets

C	Mn	P	S	Ti
0.02	0.25	0.02	0.02	0.30
Yield stress (MPa)		Ultimate tensile stress (MPa)	Elongation (%)	
170		270–330	41	

Table 2 Spectrometric analysis and the mechanical properties of galvanized 1315 IF steel sheets

6	Si	Mn	P	S	Ti
0.12	0.50	0.60	0.10	0.045	0.30
Yield stress (MPa)		Ultimate tensile stress (MPa)		Elongation (%)	
120–180		260–350		39	

Table 3 The chemical composition of filler material used during brazing operations

Chemical composition	Al	Mn	Fe	Ni	Sn	Cu
wt%	8.05	0.4	0.18	0.08	0.45	Balance

with filler electrodes. All MIG-brazing operations were prepared in an inverter-based FLEX 3000 C welding machine which was programmed using the integrated MigaCARD which holds both program data and machine software. In addition, to adjust and fix the welding speed, another machine (MX-400) was used during the brazing studies. All specimens and filler metal were cleaned with alcohol and acetone sequentially prior to brazing.

The gas metal arc brazing operations were performed in 40, 45, 50, 55, 60, 65 and 70 A welding currents in 12 V.

3 Results and Discussion

Optical microstructures of MIG-brazed steel samples were given in Fig. 1. The sectioned MIG-brazed surfaces of samples were ground on SiC paper to a 2000-grit finish and polished by diamond suspension of 0.3 μm particle size, and then ultrasonically cleaned in an acetone bath. The metallographic characterization of base metal is performed by using 2 % Nital etchant. The microstructure consists of equiaxed coarse grains with an average grain size of 18 μm. Grain size distribution is not homogeneous throughout specimen, and some finer grains are evident between the coarse grains. These findings are similar with the study of Saray et al. [1] who measured the mean grain size as 15 μm in interstitial-free steels.

In Fig. 2a, macro-structures of brazed specimens in butt joint form were given. Three characteristic regions were seen in joint zone as reported by other authors [9]

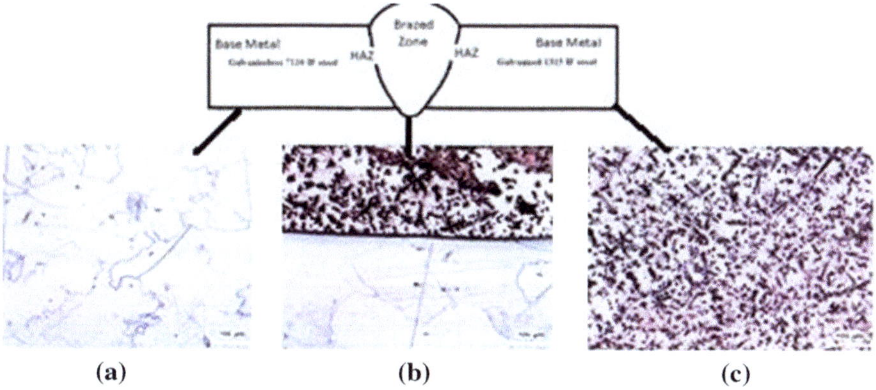

Fig. 1 Microstructures of MIG-brazed galvanizeless 7116 IF versus galvanized 1315 IF steel samples joined in butt form. This joint was performed in 55 A brazing current and 11.7 V and seemed to be the optimum parameter giving homogeneous grain size distribution

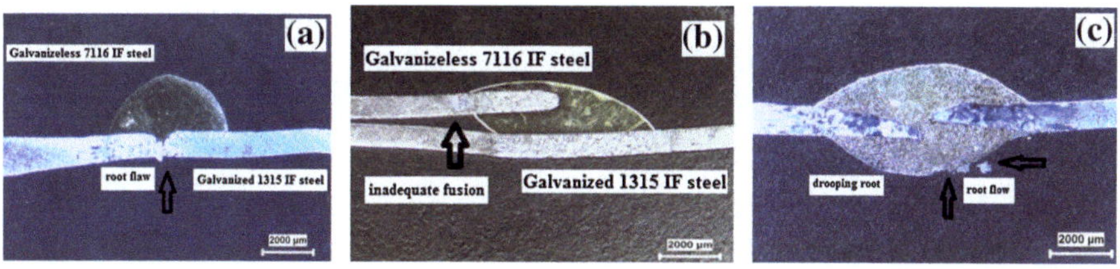

Fig. 2 Macro-structures of galvanizeless 7116 IF versus galvanized 1315 IF steel samples MIG-brazed for 5 s arc time at **a** 40 A in butt joint, **b** 45 A in overlap joint, and **c** 70 A in butt joint

as weld bead, fused and oxidized and the last region, further away from the weld bead, maintains the original appearance of the IF-steel, and has not been affected by the action of heat from the electric brazing arc [9]. The root flaws were detected at the bottom region (root) of butt joint. Similarly, inadequate diffusion was seen for the overlapped joint of IF-steel specimens joined at 45 A (Fig. 2b). Other specimens' transverse sections through MIG-brazed joints with different gap widths were seen in Fig. 2c. The brazing metal has been flowed deep into the gaps in overlapped joint regions. The brazing metal has good wettability and adherence to the base metal, covering the outer surfaces of the sheets, above and below. Clear evidence of localized fusion is observed on one of the edges of the base metal, regions where the electric brazing arc was probably concentrated. Rangel et al. has determined the similar findings [9].

In Fig. 3, Secondary electron image obtained in scanning electron microscope (SEM) microstructure of IF steel samples MIG-brazed at 55 A in overlap joint form and the EDS analysis of points from 1 to 3 were given. Here, Point-1 is in the oxygen rich zone of galvanized 1315 IF steel sheet. Some regions of specimens were exposed to oxidation. There is an oxide particle in the center of brazing region

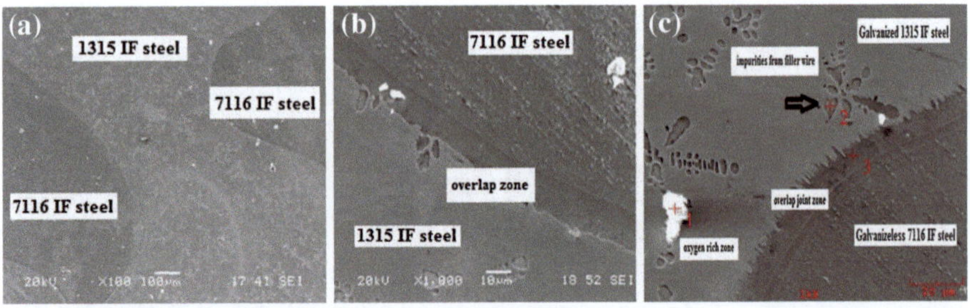

Point	Element (wt %)											
	C	O	Na	Cl	Al	P	S	Ti	Mn	Fe	Cu	Sn
1	34.545	39.658	5.530	3.405	1.778	0.131	0.571	0.076	8.608	0.607	2.344	0.949
2	-	-	-	-	11.446	0.443	-	0.183	0.331	73.805	13.136	0.041
3	0.549	-	-	-	8.450	0.364	0.353	0.077	0.048	78.995	10.235	0.403

Fig. 3 SEM microstructure of galvanized 1315 IF steel versus galvanizeless 7116 IF steel brazed in overlap joint and EDS analysis

and marked as Point-1. Its EDS analysis showed 39.65 wt% O. Point-2 and 3 contain the impurities gained from braze filler wire with the 10.235 wt% Cu and 8.450 wt% Al composition.

4 Conclusion

Cu-based CuAl8 MIG-brazing filler wire material was used. The use of Cu-base filler material leads to a reduced heat input, compared to currently apply welding processes, which may result in benefits concerning distortion, post processing and tensile strength of the joint. High feed rates prevent significant diffusion of copper into the base material. Microscopic analysis of the cross-sections showed no melting of the base material. Furthermore no penetration of braze material into the base metal along the grain boundaries at the interface neither to the lower sheet, nor to the upper sheet could be detected in overlapped joint combinations. 55 A MIG-brazing current for butt joint form and 60 A for overlap joint form seem to be the best alternatives in order to obtain good microstructure. Increase of current from these values leads to excessive fusion in both base metals and inadequate wetting of filler brazing wire. Root flaws, drooping root and micro-cracks in root of joint zone were detected in macrostructures of joints brazed in high ampere values.

References

1. O. Saray, G. Purcek, I. Karaman, T. Neindorf, H.J. Maier, Equal-channel angular sheet extrusion of interstitial-free (if) steel: microstructural evolution and mechanical properties. Mater. Sci. Eng. A **528**, 6573–6583 (2011)

2. S. Kundu, D. Roy, R. Bhola, D. Bhattacharjee, B. Mishra, S. Chatterjee, Microstructure and tensile strength of friction stir welded joints between interstitial free steel and commercially pure aluminium. Mater. Design **50**, 370–375 (2013)
3. A. Khosravi, A. Halvaee, M.H. Hasannia, Weldability of electrogalvanized versus galvanized interstitial free steel sheets by resistance seam welding. Mater. Design **44**, 90–98 (2013)
4. F. Hayat, Resistance spot weldability of dissimilar materials: BH180-AISI304L steels and BH180-IF7123 steels. J. Mater. Sci. Technol. **27**(11), 1047–1058 (2011)
5. G. Mukhopadhyay, S. Bhattacharya, K.K. Ray, Strength assessment of spot-welded sheets of interstitial free steels. J. Mater. Proc. Technol. **209**, 1995–2007 (2009)
6. S.K. Panda, D.R. Kumar, H. Kumar, A.K. Nath, Characterization of tensile properties of tailor welded IF steel sheets and their formability in stretch forming. J. Mater. Proc. Technol. **183**, 321–332 (2007)
7. G. Qin, Z. Lei, Y. Su, B. Fu, X. Meng, S. Lin, Large spot laser assisted GMA brazing-fusion welding of aluminum alloy to galvanized steel. J. Mater. Proc. Technol. **214**, 2684–2692 (2014)
8. D. Iordachescu, L. Quintino, R. Miranda, G. Pimenta, Influence of shielding gases and process parameters on metal transfer and bead shape in MIG brazed joints of the thin zinc coated steel plates. Mater. Design **27**, 381–390 (2006)
9. A.F. Rangel, L.A. Matlakhova, R.P. Darochaparanhos, A.N. Matlakhov, Evaluation of MIG-brazing welded joint by means of metallographic techniques. Weld. Int. **20**(11), 889–893 (2006)

An ESEM/EDX Methodology for the Study of Additive Adsorption at the Polymer-Air Interface

Omar Islam, Clive R. Siviour and Kalin I. Dragnevski

Abstract A methodology, combining the use of Environmental Scanning Electron Microscopy (ESEM) and Energy Dispersive X-ray (EDX) analysis was successfully developed and used to study the surface composition of thin polymer films containing adsorbed amides. It was found that the intensity of the Oxygen present in the form of amides on the surfaces of the films is dependent upon the operating conditions (e.g. gas, pressure, voltage). The latter were successfully optimised through experimentation and Monte-Carlo simulations, thus generating a 'recipe' for achieving consistent and reproducible data. Elemental X-ray maps were also successfully generated for the first time ever from surface and cross-sectional areas of the studied polymer systems, which not only allowed determination of the through-thickness distribution of the adsorbed material, but also demonstrated that the ESEM/EDX approach can be used not only to determine the overall composition of the specimen, but is also surface sensitive.

1 Introduction

For many years, adsorbed molecular layers, based on amides, have been the subject of a number of Academic problems including detergency and colloidal stability. Amides are also commercially very important additives for their surface properties at the polymer-air interface [1]. Pure polymers can often be very 'sticky' leading to problem—many will be familiar with the problem of blocking of plastic bags. To prevent this tack, amides are added to the polymer prior to formation of the polymer sheet. Once formed the amides are believed to migrate through the bulk polymer to the surface with a consequent reduction in tack and friction coefficient. However, there is only very poor composition data and essentially no structural data on these

O. Islam · C.R. Siviour · K.I. Dragnevski (✉)
Laboratory for In-situ Microscopy & Analysis, Department of Engineering Science, University of Oxford, Parks Road, Oxford OX1 3PJ, UK
e-mail: kalin.dragnevski@eng.ox.ac.uk

E.K. Polychroniadis et al. (eds.), *2nd International Multidisciplinary Microscopy and Microanalysis Congress*, Springer Proceedings in Physics 164,
DOI 10.1007/978-3-319-16919-4_19

materials to confirm this picture. In addition, some amides also show a number of unexpected effects such as an initial fall in the friction coefficient with time but followed by a subsequent rise. This behaviour has been suggested to arise from in-plane surface ordering or oxidation of the amides to a non-slip species with time [2]. Also, given that the polymer is hydrophobic, as is the 'air', the amide could easily adsorb as a monolayer or a bilayer. The use of amides as slip agents is particularly important for polymers such as polyethylene and polypropylene. Undoubtedly the key to understanding and ultimately controlling their behaviour is the acquisition of fundamental data on composition, structure and properties. However, the adsorbed layers only represent a small fraction of the total material present and hence are difficult to study. Up to date a variety of techniques have been used in an attempt to characterise these layers including X-ray powder diffraction, Quasi-Elastic Neutron Scattering and NMR [3, 4]. In a more recent study [5] we reported on the novel use of Environmental Scanning Electron Microscopy/Energy Dispersive X-ray approach, where it was clearly shown that the Oxygen present in the amide additives can be detected at certain operating conditions. It was also demonstrated that the amount of amide at the surface increases with time from an initially blown polymer film. In this study, we further investigate the 'robustness' of the ESEM/EDX approach with the aim of improving its reliability and ultimately turning this into a quantifiable surface measurement technique.

2 Materials and Methods

The polymer films containing the amide additives were obtained from Croda Ltd. The analysis was carried out on a Carl Zeiss Evo LS15 ESEM, equipped with an Oxford Instruments INCA EDX system. The effects of two main variables were extensively investigated in this study: chamber pressure and beam voltage. The method implemented was as follows. A sample of Crodamide treated film was placed on a stub and positioned within the microscope's specimen chamber. The relevant variables were then adjusted and an EDX spectrum was then obtained. Various plots and images were generated and then analysed to determine the best experimental setup for producing element maps of the film under investigation.

3 Results and Discussion

3.1 Pressure Studies

The first set of experimental results presented and discussed here investigates the effect of pressure on the Oxygen levels detected from the amide specimens. For these specimens were held at a certain chamber pressure (ranging from 10 to 200 Pa) and had multiple EDX readings taken over a period of 15–30 min to assess the dynamic effects of pressure on the polymer's Oxygen intensity. Each EDX

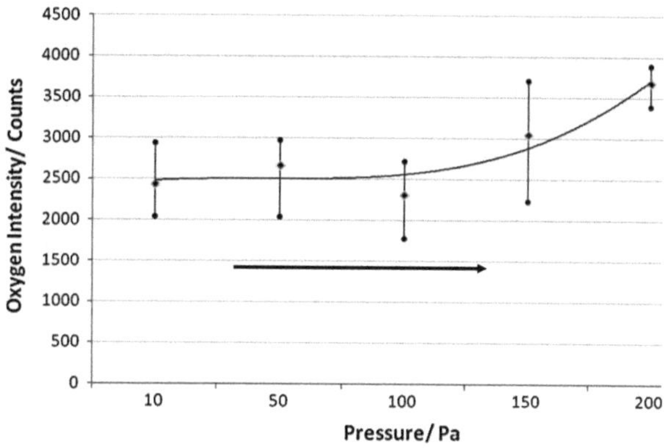

Fig. 1 Mean Oxygen intensity as a function of pressure

reading used a different but similarly sized scan area so the effects of beam damage were negligible (Fig. 1).

From the data presented in the figure above, it is evident that the Oxygen intensity increases with increasing chamber pressure. Similar trends were observed for different gaseous environments. This is most probably due to the beam skirting effect, which is common in ESEM [6, 7]. From the graph, it is also evident that there is some scatter in the data obtained from the measurements carried out at different pressures. As the readings were taken over an extended period of time by scanning new parts of the specimen each time (approximately 20 scans at each pressure), it would also appear that the amide is not distributed homogeneously on the surface of the polymer film. It is also interesting to note that at low pressures; the Oxygen intensity did not alter significantly over time, so the suggested 'sucking out' of Oxygen from the sample seems unlikely under ESEM conditions.

3.2 Voltage Studies

Another major consideration in ESEM and EDX analysis is the electron accelerating voltage of the primary beam. On the one hand, voltages that are too low may mean that electrons do not penetrate deep enough to generate representative X-ray signals and on the other, an excessive value could cause beam damage and charging artefacts. This section considers the results from the voltage experiments carried out in this study.

3.3 Monte-Carlo Simulations

Prior to the execution of the physical tests in the ESEM, and in order to obtain a better understanding of the beam-specimen interaction, simulations were performed

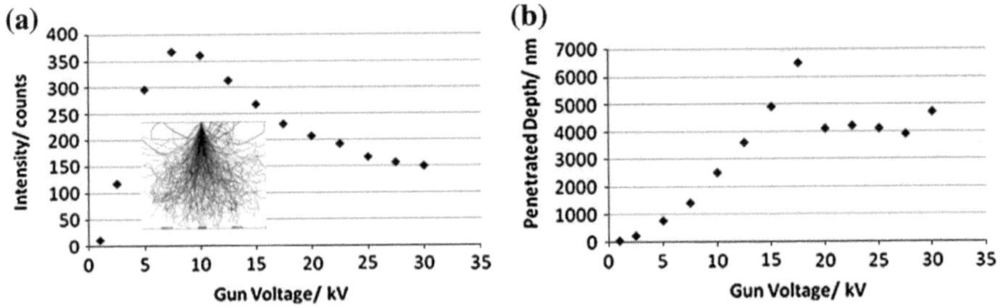

Fig. 2 Monte-Carlo simulations of Oxygen intensities (**a**) and penetration depths (**b**) in an amide-like polymer against gun voltages (*insert* example electron path)

using the Monte-Carlo based software package 'Casino v2.42'. The simulations differed from the actual experiments in that they used a high vacuum condition on a homogeneous sample of polymer containing an Oxygen compound to simulate the Crodamide. Figure 2 (insert) illustrates the electrons paths generated to form the interaction volume of the primary beam with the polymer under simulation. In addition to this visual aid, the software is able to predict the Oxygen intensities and beam penetration depths expected for different accelerating voltages (Fig. 2a, b). From the data presented in Fig. 2a, it can be observed that the Oxygen intensities increase with increasing voltage up to 7.5 kV and then drop gradually thereafter. This may be explained with the ionisations energies required to remove electrons from Oxygen atoms in the polymer (a mechanism included in the simulation which utilises the Casnati model, [8]).

As Oxygen has two electron shells, the inner K shell electrons possess a higher binding energy than the outer L shell. At higher gun voltages, there may not be as many oxygen atoms with electrons present in their outer shells to produce the characteristic X-rays [9]. This would explain why at higher gun voltages the EDX intensities begin to drop off. The simulated penetration depths in Fig. 2b appear to increase until a voltage of 17.5 kV. There is then a relatively constant penetration depth of approximately 4000 nm up to 30 kV. The constant penetration depth could be accounted for by electrons from the primary beam replacing electrons that have been ejected from oxygen atoms closer to the material surface.

3.4 Gun Voltage Testing

The following set of experiments was carried out to investigate the effects of gun voltage on the Oxygen intensities generated through the EDX analysis and compare these to the Monte-Carlo simulations.

The data shown in the figure above has some similarities with the simulations generated on the Casino software package. Although Oxygen was not detected until a gun voltage of around 10 kV, the intensity does increase in a similar manner up to

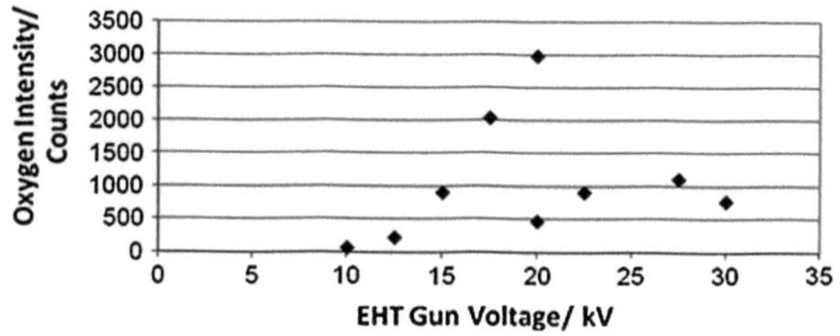

Fig. 3 Oxygen intensity versus gun voltage in an air environment

19 kV. Oxygen X-rays may not be detected below a 10 kV gun voltage as the electrons in the beam may not have sufficient energy to liberate X-rays from the sample or the X-rays generated are below the detection threshold of the EDX system. Another feature the plot is how a large sudden drop in Oxygen counts occurs at a voltage of 20 kV. Since Fig. 3 has a similar shape to the penetration depth simulation (Fig. 2b), it may be that the reduced penetration depth means that more energy is being deposited into the surface layer causing more beam damage. As beam damage and surface charging were observed at voltages above 20 kV, this would therefore explain why the subsequent points continue to be low and randomly dispersed.

3.5 Elemental X-Ray Mapping

The final part of this study, details the results obtained from the X-ray mapping experiments. Having investigated the effects of a range of pressures and gun voltages, it was decided that the most suitable settings for the mapping experiments were as follows; a 15 kV gun voltage, 70–100 Pa chamber pressure with a Nitrogen gas atmosphere, microscope working distance of 10 mm and an EDX down time of 40–43 % achievable by adjusting the beam spot size. A minimum of 50 EDX frames were captured in order to produce the individual maps. With these settings it was possible to generate reproducible elemental maps with the EDX software. X-ray maps revealing the composition of an amide treated polymer film are shown in Fig. 4a–c. From the elemental maps it is evident that the Crodamide additive is indeed present at the surface of the film. It can also be observed that the Crodamide distribution over the film's surface is largely homogeneous; although there are small localised areas showing lack of Oxygen (amide) and vice versa. This would further explain the scatter in the data that was observed in the pressure studies experiments.

Having confirmed the presence of Crodamide on the polymer surface, the next step was to attempt and identify the through-thickness distribution of the additive in the film. This was achieved by obtaining more element maps, this time along the edge of the film (i.e. cross section). The film was cut shortly before edge analysis so that the Crodamide did not have sufficient time to create a surface layer on the edge

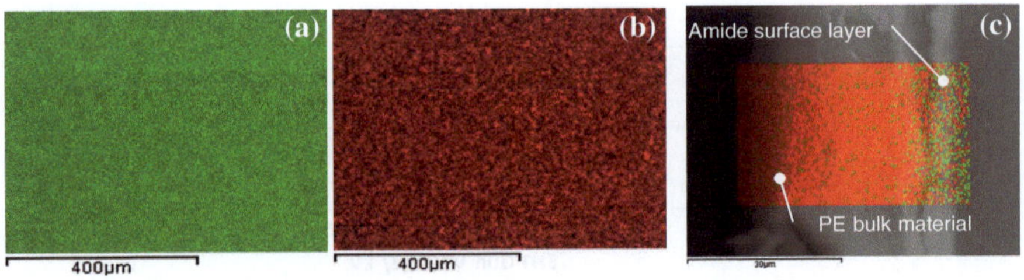

Fig. 4 Surface Carbon (**a**) and Oxygen maps (**b**) and **c** cross-sectional Carbon and Oxygen elemental maps laid over the original ESEM image (Oxygen = *green*, Carbon = *red*)

of interest. A typical example from the maps obtained from this experiment is presented in Fig. 4c. From the cross-sectional image and elemental maps shown in the figure above the distribution of the 2 main elements can clearly be seen. It is evident that the amide, here represented in the form of Oxygen (green) has migrated to the surface of the polymer film. From the cross-sectional maps it is also possible to approximately determine the thickness of the amide layer which is in the region of 4 microns, which itself is consistent with the Monte-Carlo simulations. There appear to be some Oxygen present away from the material surface, which indicates that not all of the Crodamide has diffused to the surface.

4 Conclusions

In this research the combined use of ESEM and EDX for the study of amide containing thin polymer films was further explored and developed in an attempt to determine the surface composition of the material. It was found that the intensity of the Oxygen present in the form of amides on the surfaces of the films increases with increasing gas pressure. This was attributed to the skirting effect, which is common in ESEM. Furthermore, through Monte-Carlo simulations and follow on testing it was also possible to access the effect of gun voltage on the EDX count and ultimately generate a 'recipe' for achieving consistent and reproducible data. By implementing this it was possible to generate elemental X-ray maps for the first time ever not only from surface, but also cross-sectional areas of the amide-containing specimens, This not only allowed determination of the through-thickness distribution of the adsorbed material, but also demonstrated that the ESEM/EDX approach can be used not only to determine the overall composition of the specimen, but is also surface sensitive.

Acknowledgments Dr Kalin Dragnevski would like to thank Dr Stuart Clarke (University of Cambridge) and Dr Adam Maltby (Croda Ltd.) for the provision of samples and their useful inputs.

References

1. M.X. Ramirez, D.E. Hirt, K.B. Walters, Relationship between erucamide surface concentration and coefficient of friction of LLDPE film. J. Vinyl Add. Tech. **11**, 9–12 (2005)
2. A.S. Rawls, D.E. Hirt, M.R. Havens, W.P. Roberts, Evaluation of surface concentration of erucamide in LLDPE films. J. Vinyl Add. Tech. **8**, 130–138 (2002)
3. M.X. Ramirez, D.E. Hirt, L.L. Wright, AFM characterisation of surface segregated erucamide and behenamide in linear low density polyethylene film. Nanoletters **2**, 9–12 (2002)
4. F. Miculescu, A study on the influence of the primary electronbeam on nanodimensional layers analysis. Dig. J. Nanomater. Biostruct. **6**(1), 307–317 (2011)
5. K.I. Dragnevski, A.M. Donald, S.M. Clarke, A. Maltby, Novel applications of ESEM and EDX for the study of molecularly thin amide monolayers on polymer films. Colloids Surf A Physiochem. Eng. Aspects **337**, 47–51 (2009)
6. D.E. Newbury, X-ray microanalysis in the variable pressure (environmental) scanning electron microscope. J. Res. Nat. Inst. Stand. Technol. **107**(6), 567–603 (2002)
7. G.D. Danilatos, R. Postle, Design and construction of an atmospheric or environmental SEM-2. Micron **14**, 41–52 (1983)
8. D.C. Joy, An introduction to monte carlo simulations. Scan. Microsc. **5**(2), 329–337 (1991)
9. E. Casnati, An empirical approach to K-shell ionisation cross section by electrons. J. Phys. B At. Mol. Phys. **15**, 155–167 (1982)

Enhanced Confocal Fluorescence Microscope Performance Using a Differential Pinhole

Rohan Kakade, John G. Walker and Andrew J. Phillips

Abstract While evaluating the performance of a microscope, one must consider not only the resolution but also SNR. The two criteria are each important to bio-imaging and interdependent such that attempts of optimising one generally tends to degrade the other. Here a novel detection scheme using an alternative pinhole geometry, a differential pinhole, is introduced. The scheme offers enhanced lateral as well as axial resolution resulting in an improved SNR-resolution tradeoff in a confocal fluorescence microscope. Up to 20 % resolution gain in either direction has been achieved and the performance enhancement has been demonstrated through computer simulations. The new detection scheme if used in conjunction with annular lens apertures could see further resolution gains.

1 Introduction

For an aberration-free optical microscope Abbe wave theory predicts that the theoretical lateral resolution to be limited to approximately half of the illumination wavelength [1] ($\sim$200 nm). In the axial direction it is limited to $\sim$500 nm [1]. Fluorescence microscopy with its ability to tag virtually every molecular protein [2] and produce quality images in a minimally invasive manner has proved to be an indispensable tool in live cell imaging. With improvements in quality of imaging optics, large numerical apertures and almost zero misalignment errors there now exist numerous super resolution techniques [3, 4] that surpass the well-known diffraction limit.

Fundamental to all these techniques is the tradeoff that exists between the two performance criteria—Signal-to-noise ratio (SNR) and resolution. The impact of one on the other is very apparent in confocal fluorescence microscopy (CFM) in

R. Kakade (✉) · J.G. Walker · A.J. Phillips
Applied Optics Group, Faculty of Engineering, The University of Nottingham, University Park, Nottingham NG72RD, UK
e-mail: eexrk7@nottingham.ac.uk

E.K. Polychroniadis et al. (eds.), *2nd International Multidisciplinary Microscopy and Microanalysis Congress*, Springer Proceedings in Physics 164,
DOI 10.1007/978-3-319-16919-4_20

which the tradeoff is of great significance as fluorophores are vulnerable to photobleaching (inability of fluorophores to fluoresce) and toxic cell damage [5]. In comparison to conventional widefield microscopes, CFMs although more expensive, more complex and harder to implement, are widely used owing to their ability to image thick samples. They achieve this by use of an exit pinhole which rejects out of focus light. For a CFM system with a given set of illumination and detection pupils the shape of the lateral point spread function is significantly influenced by the pinhole geometry (see 1). A smaller pinhole improves resolution of a CFM by rejecting out of focus light but at the expense of detecting fewer photons (lower SNR).

Here we investigate the use of an alternative pinhole geometry in CFM involving subtraction of the central part of the pinhole from the annular region that would provide a better SNR-resolution trade-off than the circular pinhole used in existing systems. The idea of subtracting the response from pinholes of different sizes has been discussed by various authors [6–8]. However, these papers involve subtraction of a large circular pinhole from a smaller one with appropriate weighting factors, here we argue the use of a circular and annular aperture is more noise efficient.

The key analytical expressions for CFM image formation are detailed in Sect. 2. Section 3 discusses the SNR-resolution tradeoff of conventional CFM systems. The new pinhole geometry—referred to as a differential pinhole is introduced in Sect. 4. In Sect. 5 presents computer simulation results demonstrating the new pinhole's ability to enhance both lateral and axial resolution. Finally a brief overview of the benefits of using this new detection scheme is presented in Sect. 6.

2 CFM Image Formation

The 3D lens PSF intensity in polar coordinates for a small lens semi-aperture angle α is given by (1) [9, 10].

$$PSF_{lens}(v,u) = \left| 2\pi \int_0^1 P(\rho) \exp\left(-\frac{ju\rho^2}{2} \right) J_0(v\rho)\rho d\rho \right|^2 \tag{1}$$

Here $v = kr \sin \alpha$ and $u = kz \sin^2 \alpha$ are optical radial and optical axial coordinates [11]. r is the radial displacement, z is the object defocus and $k = 2\pi/\lambda$. We assume throughout that the excitation and emission wavelengths to be equal. Here $\sin \alpha \approx a/f$ is the numerical aperture of either lens. a is the pupil radius, f is the focal length (equal for either lens under unity magnification), α is the lens semi-aperture angle and λ is the illumination wavelength.

For a CFM system with the objective and detection lens PSFs defined by PSF_1 and PSF_2 respectively and a detector pinhole with transmittance $Pin(v)$ the 3D image formed (assuming that the excitation and emission wavelengths to be equal) is given by (2) [12]. $O(v, u)$ is the object fluorophore concentration and (*) represents 3D convolution.

$$I(v,u) = O(v,u) * (PSF_1(v,u)(PSF_2(v,u) * Pin(v))) \tag{2}$$

3 SNR-Resolution Trade-Off

Resolution based on the Rayleigh criterion is found by computing the full-width half maximum (FWHM) of the PSF (viz., the system response to a point object) along the lateral and axial directions. Amongst the various noise sources that degrade the signal quality, shot noise has been acknowledged to be the dominant noise contributor in high-quality bio-imaging [13]. It is well known that the photon distribution follows poissonian characteristics. Thus, the mean and variance of the distribution are equal and the SNR can be readily calculated as the ratio of mean (viz., the sum of detected photons at the image plane) to standard deviation. This can be deduced to be the square root of the total detected photons N at the image plane.

SNR versus resolution as a function of pinhole radius in both axial and lateral directions has been plotted in Fig. 1. The SNR has been normalised by the number of photons in the illumination and the resolution (FWHM) has been normalised by NA/λ in the lateral and NA^2/λ in the axial directions. The plot shows that applications demanding a high SNR would require a large pinhole and this would invariably impede resolution.

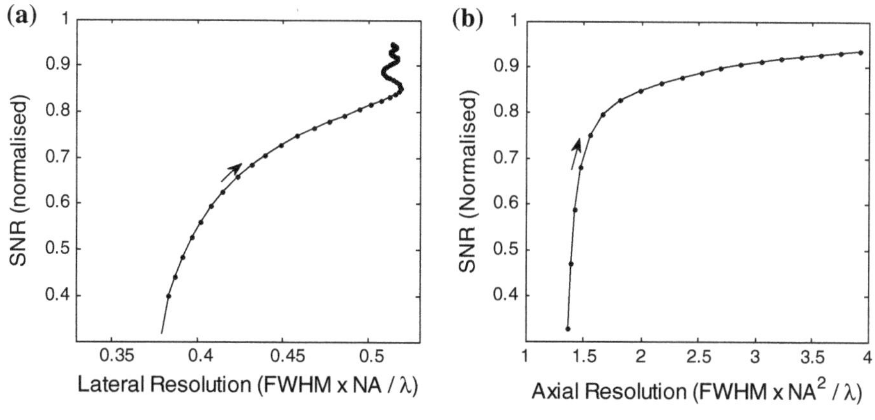

Fig. 1 Variation of SNR and resolution (lateral (**a**) and axial (**b**)) as a function of pinhole radius for a confocal fluorescence microscope. *Arrows* indicate the direction of increasing pinhole radius

As can be seen from (2) the pinhole function directly affects the shape of the 3D PSF and experimenting with other pinhole shapes might offer a better SNR-resolution trade-off. In the next section we introduce the concept of a differential pinhole—an alternative pinhole arrangement that brings this about.

4 Differential Pinhole

Consider a new pinhole geometry that consists of two concentric annular regions defined mathematically as shown in Fig. 2. These two regions direct parts of the incoming light onto separate detectors. In Fig. 2 and throughout this paper r_{pin} is the outer pinhole radius and r_{inner} is the radius of the inner transparent annuli. The tilt of the pinhole plane viz., the angle between the plane and the optical axis, does not significantly impact the system performance as we assume detectors, which are large compared the pinhole, to be placed very close to the pinhole (as in most CFM systems). If N_+ and N_- denote the number of photons collected by detector D1 and D2 respectively then the received signal S is given by $N_+ - N_-$ and the total noise by the sum of the noise generated by D1 and D2.

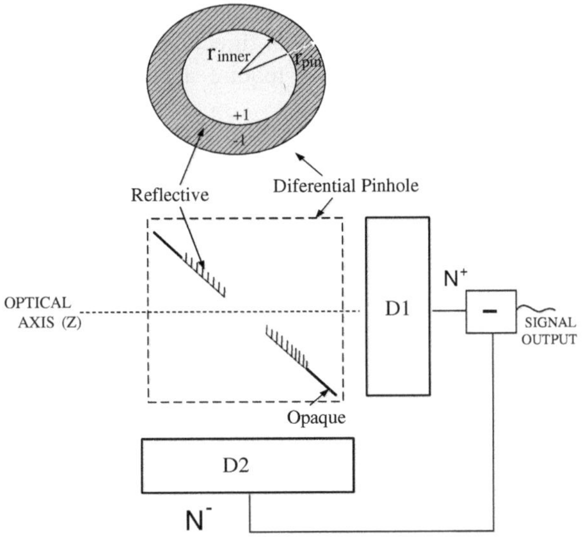

Fig. 2 Detector arrangement showing how a differential pinhole is realised using two detectors D1 and D2. The pinhole's projection on a plane perpendicular to the optical axis is shown (*top*). D1 collects the photons incident within a pinhole radius r_{inner} while D2 detects photons reflected within the region $r_{inner} < r < r_{pin}$. The photon intensities are subtracted to produce the final signal

5 Improved Performance

The lateral/axial resolution and the SNR of the differential pinhole as a function of pinhole radius (i.e. the pinhole outer radius r_{pin}) can be plotted as shown in Fig. 3a, c. Although there is an overall reduction in SNR using the differential pinhole, there is a significant performance gain realised through the reduction in PSF half-width (see Fig. 4).

A constraint $r_{inner} = r_{pin}/\sqrt{2}$ has been enforced to ensure that the inner and outer annuli of the differential pinhole have equal area as this offers a good SNR-resolution trade-off. This is because if $r_{inner} < r_{pin}/\sqrt{2}$, the SNR drops with D2 detecting an increasing number of photons. And if $r_{inner} > r_{pin}/\sqrt{2}$, the performance mimics that of a circular pinhole system i.e. offering no resolution gain. Curves in Fig. 3b, d, plotted under the above constraint, indicate a wide range of pinhole radii (0.2–0.8 λ/NA) for which the differential pinhole based system offers improved performance in comparison to the circular pinhole based system. become equal for the two systems at these radii (see Fig. 3a, c). The same follows for the axial response.

At larger pinhole radii ($r_{pin} \sim > 1\ \lambda/\text{NA}$) both the circular and differential pinhole based CFM systems offer the same lateral resolution as the FWHM tend to become equal for the two systems at these radii. As well as the form of the SNR

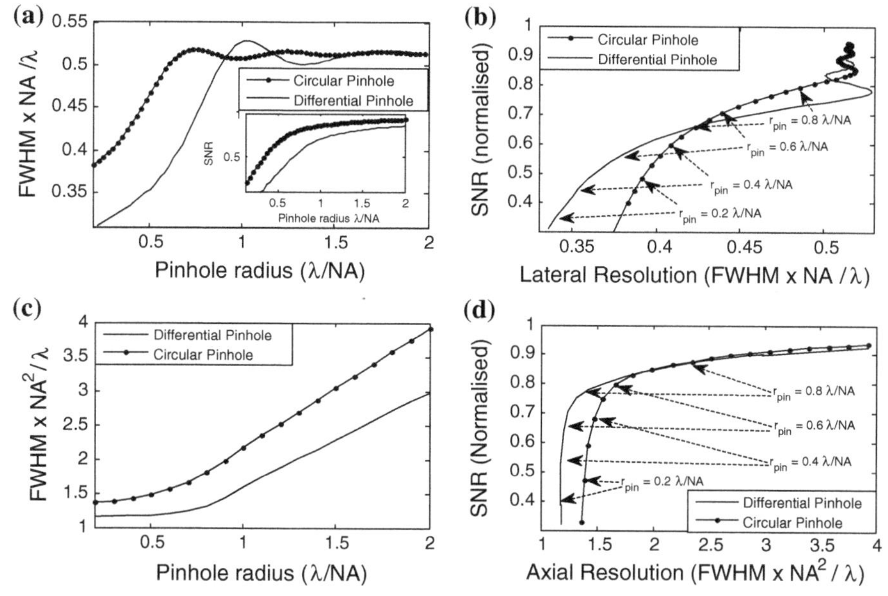

Fig. 3 **a** Lateral and **c** axial resolution characteristics of a differential pinhole and a circular pinhole for a CFM system. **b** Lateral and **d** axial SNR versus resolution characteristics. Improved performance is indicated by a *curve* nearer the *top left* in plot (**b**) and (**d**). The *arrows* locate the trade-off point for a given outer pinhole radius for both the systems

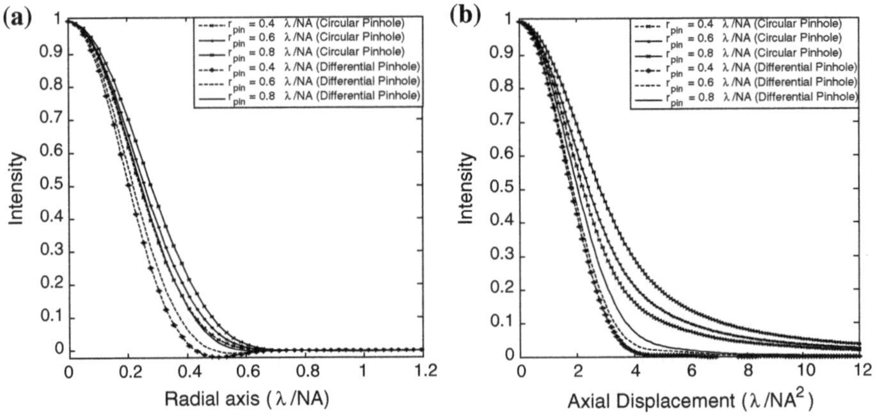

Fig. 4 Normalised lateral (**a**) and axial (**b**) confocal responses at pinhole radii 0.4, 0.6 and 0.8 λ/NA showing the reduction in PSF half width brought about by the differential pinhole system

resolution characteristic, it is also important to note that the PSF's produced by the differential pinhole do not have significantly enhanced sidelobes or other detrimental characteristics (see Fig. 4). The most attractive feature of CFM system has always been the large axial resolution it offers, allowing sectioned imaging of thick specimens. As shown in Fig. 4b the differential pinhole produces a much smaller PSF half-width, although at the cost of reduction in SNR.

6 Conclusions

We observe an approximately 20 % decrease in the PSF mainlobe width in either direction with nearly non-existent sidelobes. In particular the enhanced axial resolution achieved makes the differential pinhole system well suited for imaging thick samples compared to circular pinhole based systems. The approach above may be also implemented using a single CCD camera in the detector plane in a similar way to the arrangement in [14]. This approach has the advantage that many different pinhole sizes and regions used for image formation may be selected from a single measurement. The advantage of the detector approach is, however, that the fast point detectors may be used which do not require readout of all pixels. With many applications in fluorescence microscopy requiring detection of very low signals, the high dynamic range of PMTs make them more suitable than most high-quality CCDs. The approach of subtracting signals from different sized pinholes [6] although offers improved performance suffers from increased noise which results in a lower SNR. By subtracting signal from different annular regions better SNR is achieved for the same level of performance gain using the differential pinhole. The new system although requires two detectors calls for no significant modification of the generic setup and can be readily used in existing systems.

References

1. L. Schermelleh, R. Heintzmann, H. Leonhardt, A guide to super-resolution fluorescence microscopy. J. Cell Biol. **190**(2), 165–175 (2010)
2. R. Heintzmann, G. Ficz, Breaking the resolution limit in light microscopy. Briefings Funct. Genomics Proteomics **5**(4), 289–301 (2006)
3. J. Beweresdorf, R. Schimdt, S.W. Hell, Comparison of I^5M and 4Pi-microscopy. J. Microsc. **222**(2), 105–117 (2006)
4. M. Fernandez-Suarez, A.Y. Ting, Fluorescent probes for super-resolving resolution imaging in living cells. Nature **9**, (2008)
5. M.G. Somekh, K. Hsu, M.C. Pitter, Effect of processing strategies on the stochastic transfer function in structured illumination microscopy. Opt. Soc. Am. **28**(9), 1925–1934 (2011)
6. R. Heintzmann, V. Sarafis, P. Munroe, J. Nailon, Q.S. Hailey, T.M. Jovin, Resolution enhancement by subtraction of confocal signals takes at different pinhole sizes. Micron **34**(6), 293–300 (2003)
7. Y. Wang, C. Kuang, Z. Gu, X. Liu, Image subtraction method for improving lateral resolution and SNR in confocal microscopy. Opt. Lasers Technol. **48**, 489–494 (2012)
8. M. Martinez-Corral, M.T. Caballero, C. Ibanez-Lopez, V. Sarafis, Optical sectioning by two pinhole confocal fluorescence microscopy. Micron **34**(6–7), 313–318 (2003)
9. M. Born, E. Wolf, *Priciples of Optics* (Cambridge University Press, Cambridge, 1993)
10. C.J.R. Sheppard, Limitations of paraxial debye approximation. Opt. Lett. **38**(7), 1074–1076 (2013)
11. M. Gu, Principles of three dimensional imaging in confocal microscopes (World Scientific, Singapore, 1996)
12. C.J.R. Sheppard, T. Wilson, Image formation in confocal scanning microscopes. Optik **55**(4), 331–342 (1980)
13. M.G. Somekh, K. Hsu, M.C. Pitter, Stochastic transfer function for structured illumination microscopy. Opt. Soc. Am. **26**(7), 1630–1637 (2009)
14. E.S. Ortiga, C.J.R. Sheppard, G. Saavedra, M. Martinez-Corral, A. Doblas, A. Calatayud, Subtractive imaging in confocal scanning microscopy using a CCD camera as a detector. Opt. Lett. **37**(7), 1280–1282 (2012)

Thermal Neutron Detection by Entrapping ^{6}LiF Nanodiamonds in Siloxane Scintillators

M. Degerlier, S. Carturan, T. Marchi, M. Dalla Palma, F. Gramegna, G. Maggioni, M. Cinausero and A. Quaranta

Abstract Exploiting the long experience in design and production of scintillating mixtures based on siloxane matrices with combinations of primary dye and waveshifter, a first set of ^{6}LiF loaded scintillator disks have been produced and the involved steps for their preparation are herein described. Preliminary results as for their light response towards thermal neutrons are reported as well. The preservation of transparency and mechanical integrity of the scintillator material is challenging when introducing the inorganic salt LiF, which is a "foreign body" to the organic polysiloxane host matrix. Different strategies such as synthesis of nanoparticles and surface functionalization have been pursued to succeed in the entrapment of the neutron converter whilst maintaining moderate light output, optical clarity and flexibility of the base scintillator.

1 Introduction

The importance of revealing nuclear weapons concealed in trucks or boats at the borders, using Radiation Portal Monitors (RPM) has grown up, unfortunately, in the last years. Furthermore, a renewed interest in neutron detection has grown up again both in nuclear physics, mainly related to the development of Radioactive Ion Beam

M. Degerlier (✉)
Science and Art Faculty Physics Department, Nevsehir Haci Bektas Veli University, Nevsehir, Turkey
e-mail: meltem.degerlier@lnl.infn.it; mdegerlier@nevsehir.edu.tr

S. Carturan · T. Marchi · M. Dalla Palma · F. Gramegna · G. Maggioni · M. Cinausero · A. Quaranta
INFN - Laboratori Nazionali di Legnaro, Legnaro, Italy

M. Dalla Palma · A. Quaranta
Department of Industrial Engineering, University of Trento, Trento, Italy

S. Carturan · G. Maggioni
Department of Physics and Astronomy, University of Padova, Padua, Italy

E.K. Polychroniadis et al. (eds.), *2nd International Multidisciplinary Microscopy and Microanalysis Congress*, Springer Proceedings in Physics 164,
DOI 10.1007/978-3-319-16919-4_21

facilities, and in applied physics related fields, such as materials analysis by neutron imaging and diffraction spectroscopy. The use of ^{3}He for the detection of moderated neutrons with almost full rejection of γ-rays generated events has become prohibitive, due to the exceedingly high price/liter of this rare isotope, though the reliability of ^{3}He-based systems is still unrivalled. This has triggered a widespread and intensive research work aimed at ^{3}He replacement. Systems based on scintillators, both inorganic [1] and organic [2–5] have been recently proposed as neutron detectors, while compact and robust silicon based detectors where neutron converter thin films are used for signal generation have been recently ideated and proved their excellent performances [6]. Siloxane-based scintillators are, in principle, very simple and economic tools for particle detection. Their capability to detect thermal neutrons by loading with ortho-carborane has been demonstrated, but the limited solubility of the organoboron compound hampered their further development [7].

Other neutron converters have been tried as dopants for siloxane, such as Gd and Cd. However, the difficulty in dissolving metal-organics inside the siloxane matrix, while preserving transparency and fluorescence intensity, the toxicity of Cd itself and the unfeasible n-γ discrimination, owing to the intrinsic emission of a flash of γ-rays following neutron capture on Gd or Cd, make those systems less attractive than ^{6}Li doped scintillators. In this case, the triton and alpha particles from the ^{6}Li reaction can differently excite the material with respect to γ-rays, thereby enabling n-γ discrimination by pulse shape analysis (PSA).

In this preliminary work, we report on the production of a “standard” siloxane scintillator, added with a suitable combination of primary dye and wavelength shifter, where ^{6}LiF nanocrystals have been embedded. The choice of LiF as lithium bearing compound is dictated by several key features, such as minimal molecular volume, higher Li mass% content as compared to other Li salts and negligible hygroscopicity, which induces opacity and, in turn, severe light output loss. The critical issue to be addressed is the addition of ^{6}LiF to the siloxane matrix while preserving the light output. Therefore, the entrapment of nanoparticles, which are known to minimize light scattering effects, has been herein exploited for the first time. The synthesis of the nanodiamonds of ^{6}LiF has been pursued by two different routes: the co-precipitation method in a water:ethanol solvent:co-solvent system [8] and the high temperature decomposition of trifluoroacetate lithium precursor (LiTFA) in oleic acid/octadecene environment [9].

2 Materials and Methods

Polysiloxane based scintillators are obtained by Pt catalyzed addition reaction, as described previously [4, 5]. In case of scintillators, one can add suitable dyes and neutron converter to the base resin, before vulcanization. In this case, 2,5-diphenyloxazole (PPO) was chosen as primary dye and Lumogen Violet (Basf, LV) as wavelength shifter.

As neutron absorber ^{6}Li was chosen because of its unique and clean decay channel consisting of a 2.73 MeV triton and 2.05 MeV alpha particle, as mentioned above. Among the various lithium compounds, lithium fluoride displays the lowest water absorption and, very recently, LiF nanoparticles synthesis has been reported [7]. The approach of nanoparticles synthesis can be advantageous since the use of nano-sized particles can minimize light scattering and increase the transparency. In this work, ^{6}LiF nanoparticles were prepared using two different methods: co-precipitation and thermal decomposition of LiTFA. In the first process, ^{6}Li chloride was prepared from hydrated ^{6}Li oxide, by treating the suitable amount in a Pyrex capsule with hot concentrated HCl acid. Hence, the white salt was dissolved in a fixed volume of water:ethanol mixture to obtain a 0.2 M solution. Then it was slowly added to an already prepared solution of NH_4F in the same concentration and volume of solvent:co-solvent. The reaction was continued for about 1 h, then the nanocrystals were collected by centrifugation.

In the other method, the preparation of ^{6}LiF with a shell of oleic acid started from LiTFA, obtained by treating hydrated ^{6}Li oxide with hot trifluoroacetic acid. Then, LiTFA was mixed with oleic acid and octadecene (1:1 mixture) and the solution was heated under argon at 320 °C for 1 h. After cooling, excess ethanol was added and the nanocrystals with oleic acid functionalization were recovered by centrifugation.

Samples of polysiloxane with added ^{6}LiF were prepared with amount of 0.5, 1 and 2 wt% (wt% of ^{6}Li) by simple dissolution of oleic acid-capped particles, while mechanical blending was applied to disperse the nanoparticles from co-precipitation (water:ethanol 0:1). The samples with LiF OA, which are about 1 mm in thickness, appeared as homogeneous, though as the amount of LiF increases the optical clarity decreases remarkably. In the sample with 1 % LiF from co-precipitation several small undispersed grains can be seen, causing scattering, while the sample with 2 % LiF is completely in-homogeneous and opaque, with some large white undissolved clusters.

3 Results and Discussion

In Fig. 1 SEM images of the powders obtained by co-precipitation are shown. When equal amount of water and ethanol is used, micron sized cubes, highly regular with a quite narrow size distribution are produced. If only ethanol is used, the particles are much smaller and their size decreases down to some hundreds of nm.

In the case of LiTFA decomposition, SEM inspection evidences flower-like structures, as visible in Fig. 2. Oleic acid may promote aggregation of nanocrystals during growth at high temperature, with formation of different shapes. Energy dispersive spectroscopy applied to a single structure reveals the predominant presence of fluorine (Fig. 2), while signals of carbon and oxygen from oleic acid are

Fig. 1 SEM images of ^{6}LiF crystallites obtained using different ratios water:ethanol (*top panel* 1:1, *bottom panel* 0:1)

Fig. 2 SEM image and EDS spectrum of ^{6}LiF oleic acid capped

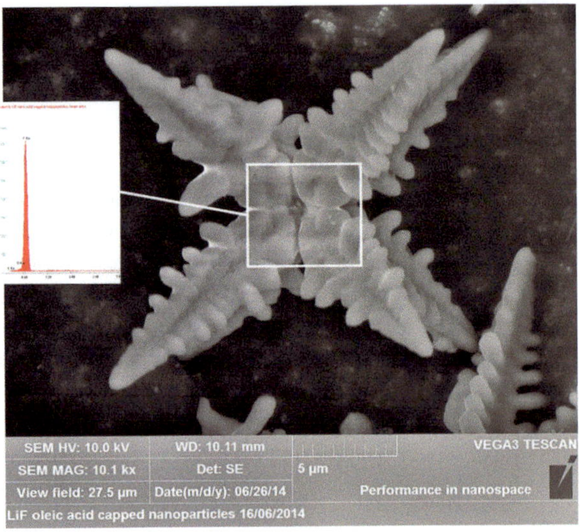

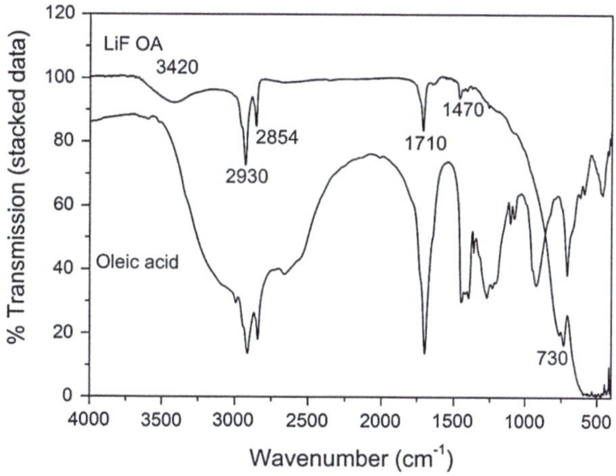

Fig. 3 Infrared spectra of pure oleic acid and LiF OA capped particles

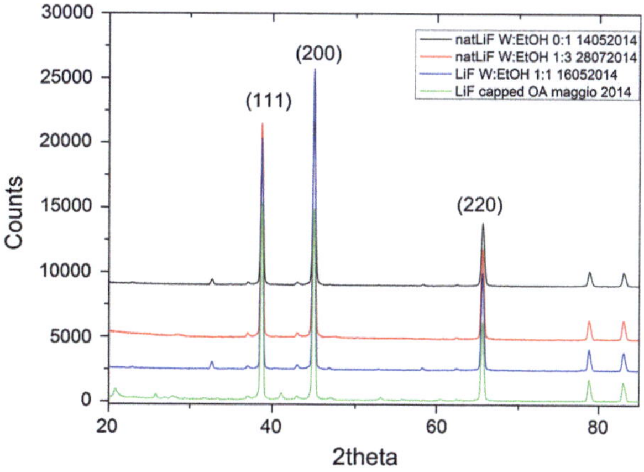

Fig. 4 HR-XRD patterns from powders synthesized by the two different pathways

very weak. The long chains of oleic acid are presumably bound to the surface of the crystals, hence the amount of C and O is low as compared to LiF.

The infrared spectra of pure oleic acid and LiF OA capped particles are shown in Fig. 3. The main features of oleic acid are present also in the LiF sample, confirming the successful capping of the inorganic crystal with the organic moiety. The disappearing of the wide band due to –OH of the carboxylic acid in the range 3200–2500 cm^{-1} confirms the formation of carboxylate anion and coordination with LiF particle surface.

High-resolution X-ray diffraction analysis has been performed on powders (Fig. 4). All the samples show the typical XRD pattern of crystalline LiF (griceite).

As for the particle size, the FWHM of the main peak (200) can be observed for a qualitative comparison. This parameter did not vary so much between the different powder samples, though a slight peak broadening is visible as the amount of water decreases, indicating a smaller crystallites size. The peak width of LiF OA sample is

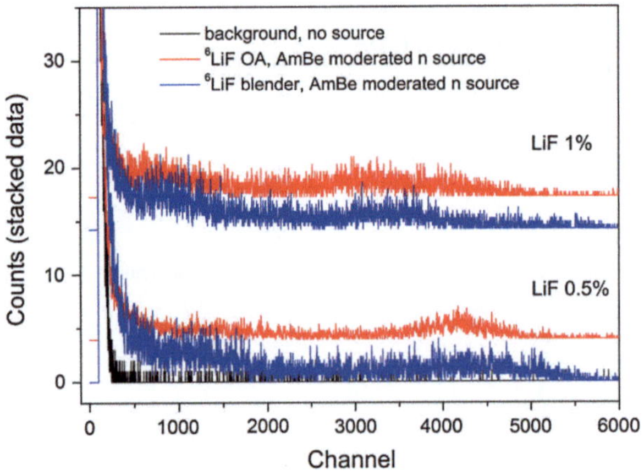

Fig. 5 Pulse-height spectra collected by exposing the samples to thermalized neutrons

quite similar to the other cases. Hence, it can be inferred that SEM images are not appropriate to estimate the crystals size, which is presumably in the nanometer scale.

Exposure to thermal neutrons, produced by using an AmBe neutron source moderated with some polyethylene bricks, allowed to observe the neutron capture peak in the region at high channels in the spectra of Fig. 5. The sample obtained by adding ^{6}LiF OA capped (0.5 wt%) seems to give a sharper signal, with better signal to noise ratio than the sample obtained with blender. Response from samples with higher content of LiF is worse, owing to decreased transparency and higher light loss. The shoulder which appears at lower channels has been ascribed to γ-rays, emitted from the source along with neutrons and not properly shielded.

4 Conclusions

^{6}LiF nanoparticles have been synthesized by co-precipitation and LiTFA thermal decomposition. In the first case, ^{6}LiF nanoparticles with variable size have been obtained controlling the ratio between solvent and co-solvent. Following thermal decomposition of LiTFA, ^{6}LiF crystals have been obtained, with a cap surface layer of oleic acid, as demonstrated by FTIR analyses. SEM-EDS analyses confirmed the synthesis of nanodiamonds by co-precipitation with size ranging from 50 to 150 nm, while micron-sized particles with flower-like shape have been observed as a result of LiTFA decomposition in oleic acid. However, XRD spectra of the powders obtained by different methods are comparable, thus indicating that also in the case of LiTFA nanocrystals have been obtained. The crystallites growth in oleic acid environment can lead to agglomeration and grafting with the long aliphatic chain may occur at the end of the process.

^{6}LiF from the two pathways have been entrapped in siloxane scintillators; both samples series proved to be sensitive to thermal neutrons and pulse-height spectra display the neutron capture peak. Further work is in progress to improve the optical transparency and to evaluate the effect of sample thickness on the response to neutrons.

References

1. C.W.E. van Eijk et al., Nucl. Instrum. Meth. A **529**, 260–267 (2004)
2. N. Zaitseva et al., Neutron detection with single crystal organic scintillators. Proc. SPIE **7449** (2009)
3. Z.W. Bell et al., Organic scintillators for neutron detection. Proc. SPIE **4784** (2003)
4. S. Carturan et al., Rad. Prot. Dosimetry, 1–6 (2010)
5. S. Carturan et al., Eur. Phys. J. FP **129**, 212 (2014)
6. M. Barbagallo et al., Rev. Sci. Instrum. **84**, 033503 (2013)
7. A. Quaranta et al., J. Non-Cryst. Solids **357**, 1921–1925 (2011)
8. N.D. Alharbi et al., J. Phys. D Appl. Phys. **46**, 035305 (2013)
9. D. Ya-Ping et al., Dalton Trans. 8574–8581 (2009)

High Temperature Reliability of Ta-Based and TiW-Based Diffusion Barriers

Nando Budhiman, Ulrich Schürmann, Björn Jensen, Steffen Chemnitz, Lorenz Kienle and Bernhard Wagner

Abstract Two types of diffusion barriers, Ta-based and TiW-based, with Ni layer on top were investigated to determine their reliability for high temperature applications up to 600 °C. Both barriers were deposited in form of stacked layer on 200 mm silicon wafers using standard production tools for physical vapor deposition. In order to investigate their reliability, the barriers have been annealed inside a vacuum chamber for 24 h up to 600 °C. Before and after annealing, three ex situ analysis techniques consisting X-ray diffraction method, 4-probe measurement method, and transmission electron microscopy (TEM) have been applied. After annealing for 24 h at 600 °C, Ni–Ta and Ni–TiW phases from Ta- and TiW-based barriers, respectively, were determined and both barriers showed a tolerable increase of sheet resistance. TEM analysis determined no diffusion of materials into the substrate using both barriers and this result was confirmed by XRD analysis.

1 Introduction

In microsystem technology, thermoelectric generators (TEGs) can be fabricated by bonding two substrates consisting p- and n-type semiconductor. The bonding is established using a bond solder having higher re-melt temperature rather than temperature, where the TEG will be applied. To deposit such solder, normally by electroplating, one requires a plating base, which serves as diffusion barrier as well as adhesion promoter. The objective of this investigation is to develop a barrier, which prevents diffusion into the substrate during application at elevated temperatures up to 600 °C while maintaining its initial adhesion and electrical conductivity.

N. Budhiman (✉) · U. Schürmann · S. Chemnitz · L. Kienle · B. Wagner
Institute of Material Science, Christian-Albrechts-Universität zu Kiel, Kiel, Germany
e-mail: nbu@tf.uni-kiel.de

B. Jensen · S. Chemnitz · B. Wagner
Fraunhofer Institute for Silicon Technology (ISiT), Itzehoe, Germany

E.K. Polychroniadis et al. (eds.), *2nd International Multidisciplinary Microscopy and Microanalysis Congress*, Springer Proceedings in Physics 164,
DOI 10.1007/978-3-319-16919-4_22

An excellent barrier's stability against high temperature have been shown from titanium-tungsten (TiW) [1] and tantalum (Ta) [2] layers. However, studies on their thermal stability have been reported only for 1 h of annealing time. Thermal expansion of the material should be minimized during barrier's application at high temperature. Titanium-nitride (TiN) exhibits a reduction in its residual stress when annealing between 450 and 800 °C for 1 h [3]. Stacked layers of a fabricated barrier could also be an advantage to enhance its reliability [4]. However no increase in resistance has been reported after high temperature annealing.

2 Reliability Investigations

2.1 Experimental Details

Ta-based and TiW-based barriers with Ni as top layer were fabricated in form of stacked layers, which consist of Ta and TiN for Ta-based barrier and TiW and TiN for TiW-based barrier, inside a class-100 clean room. They were deposited on standard 200 mm silicon wafers using standard production tool for physical vapor deposition. Figure 1 illustrates the cross-section of both barriers after fabrication.

Three methods have been performed before and after annealing to investigate the barrier's reliability. The annealing was performed under vacuum from 200 to 600 °C with steps of 100 °C and a holding time of 24 h for each temperature. The first applied method was X-ray diffraction analysis (XRD) using θ–2θ method (using Cu-Kα radiation operated at 40 kV and 40 mA) to identify the phases formed during annealing. Two different slit widths (2 and 1 mm for source and detector, respectively) were used. Scans were performed from 25° to 65° 2θ angle steps of 0.05° and a step time of 5 s. Transmission electron microscopy (TEM) was applied as second method which was performed using analytical TEM FEI Tecnai F30 STwin microscope (300 kV, field emission gun (FEG) cathode, spherical aberration coefficient $C_s = 1.2$ mm) to determine the possible diffusion into the substrate. For this particular investigation the cross sectional samples were prepared by the sandwich method including ion beam milling (precision ion polishing system Gaten PIPS 691). The composition was determined using energy-dispersive X-ray spectrometry (EDX) (Si(Li) EDAX detector) in the HAADF-STEM mode. The energy filtered TEM (EFTEM) has been performed to determine nitrogen diffusion. All methods of TEM investigation has been performed on un-annealed and 24 h-annealed barriers at 600 °C. As third method, 4-point probe measurement was performed on all samples to measure the sheet resistance of the barriers.

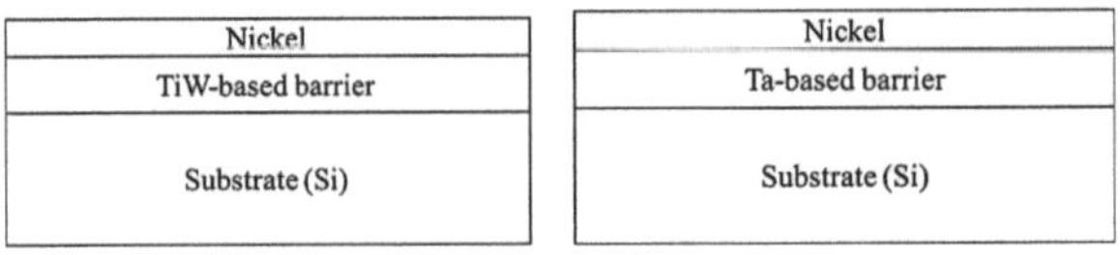

Fig. 1 Sketch of cross-section of TiW-based (*left*) and Ta-based (*right*) barriers

2.2 XRD Analysis

Two plots of XRD spectra for TiW-based and Ta-based barriers are shown in Fig. 2. The peaks of all elements from both barriers are present [1, 2, 5–7]. New phases, which are indicated by a peak at 44° in TiW-based barrier and peaks at 27.5° and 42.8° in Ta-based barrier, were determined after annealing up to 600 and 500 °C for TiW- and Ta-based barriers, respectively. These new peaks or phases could be originated from the compounds between Ni and TiW, and Ni and Ta.

2.3 TEM Analysis

The TEM investigation consisted of EDX elemental mapping, EDX line and point scans, and EFTEM analysis. These four methods have been performed to both barriers before and after annealing, except EDX point scan, which has been performed after annealing. EFTEM analysis has been performed before and after annealing only to Ta-based barrier. Ni diffusion was only determined into the top part of the barrier. This supports the XRD result as well. The EFTEM analysis revealed that no nitrogen diffusion into Ta adjacent layers has been taken place.

2.4 Four-Probe Measurement

As depicted in Fig. 3, both barriers exhibited approx. 130 % increase of sheet resistance after annealing at 600 °C. Moreover, they were still measurable even

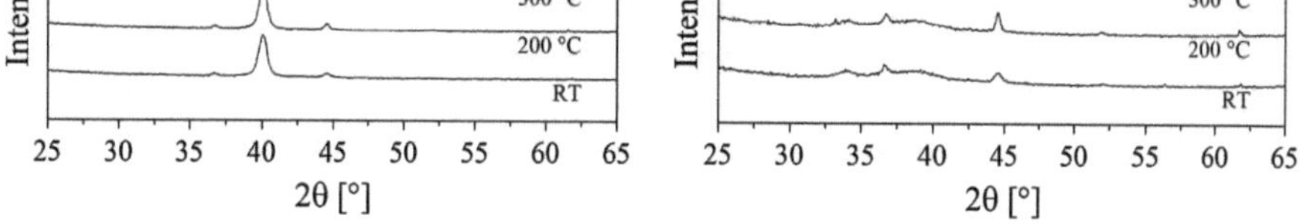

Fig. 2 The XRD spectra of TiW-based (*left*) and Ta-based (*right*) barriers after annealing at given temperatures

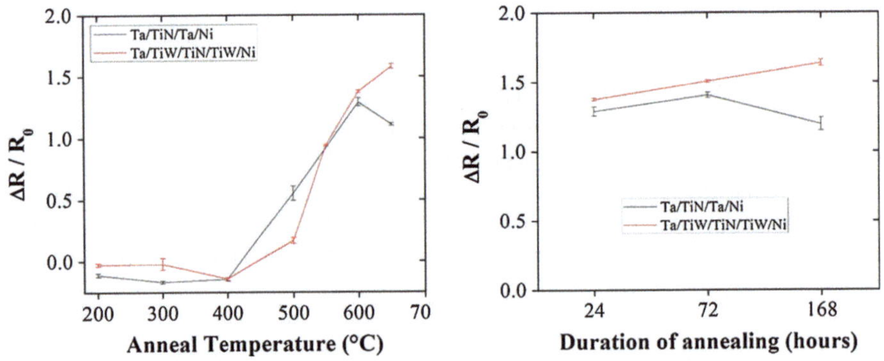

Fig. 3 Plot of increase in sheet resistance of each barrier, measured after annealing at given temperatures (*left*) and for given duration (*right*)

after annealing at 650 °C. After 72 h of annealing at 600 °C, their sheet resistance increased slightly in contrast to 24 h of annealing. After 168 h, a 7 days annealing period, TiW-based and Ta-based barriers exhibited a slightly increase and slightly decrease of sheet resistance, respectively. This investigation concludes that both barriers exhibits a tolerable increase in sheet resistance, proving that they can still maintain their electrical conductivity during high temperature application.

3 Diffusion Barriers for Thermoelectric Generators

A test structure for investigating the wafer bonding in TEG fabrication has been developed. A lithography process has been performed on two pairs of wafer, one pair with TiW-based barrier and the other with Ta-based barrier, in order to create galvanic structures, which was based on TEG structure pattern. The galvanic structure was made from Ni–Sn electroplating and was served as bond solder for establishing the wafer bonding (see Fig. 4 to illustrate the barrier, the galvanic structure, and TEG structure before wafer bonding). After bonding, the wafers were diced to get single dies of TEG test structure with size of 5 × 4 mm^2 each.

To investigate the bond strength and barrier adhesion respectively, shear test has been performed on test structure dies; 18 dies from un-annealed dies and 20 dies from 600-°C-24 h-annealed dies, from each bonded wafer using a dedicated tool. Table 1 summarizes the shear test results and statistic numbers of break location. A high bond strength was shown before and after annealing and most of the break was occurred within Si substrate. This concludes that the barrier adhesion was high and it was more than the strength for breaking the substrate.

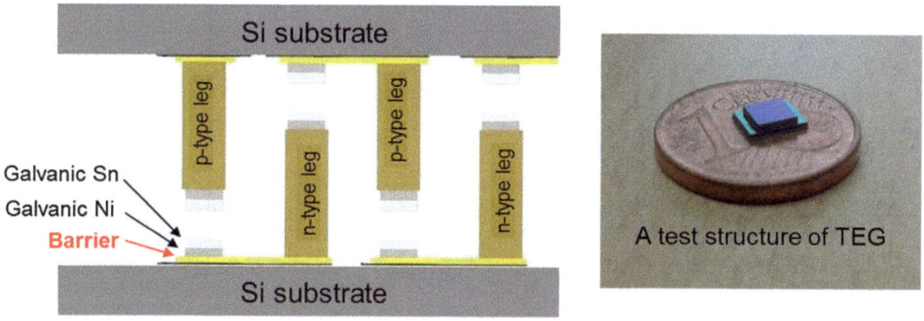

Fig. 4 A sketch of TEG structure showing where the barrier and galvanic structure are deposited (*left*) and a fabricated test structure (*right*)

Table 1 Shear test result from un-annealed and 600-°C-annealed test structure with the statistical assessment of break locations

			Statistic of break location due to shear test (%)				
Annealing		Bond strength (MPa)	Substrate breaks	Oxide delaminated	Platingbase delaminated	Ni–Sn delaminated	Within Ni–Sn
Before	MV	102.1	68.8	0	18.4	0.8	11.9
	SD	25.2	28.4	0	18.5	1.6	13.8
After	MV	138.8	81.4	0	11.9	0	7.8
	SD	21.6	10.7	0	7.2	0	8.3

(*MV* mean value, *SD* standard deviation)

4 Summary

The reliability of TiW-based and Ta-based barriers for high temperature application up to 600 °C has been investigated. It was determined that the barrier could prevent diffusion into the substrate, indicating their potential use in TEG for protecting the thermoelectric element from diffusion. It was also shown that the barriers exhibit a good quality adhesion on Si substrate and a tolerable increase of sheet resistance. The later was due to alloying between Ni and TiW in TiW-based barrier and between Ni and Ta in Ta-based barrier after annealing. This investigation shows that both barriers could be reliable and applicable for high temperature application, especially thermoelectric generator.

Acknowledgments This investigation was supported by The Federal Ministry of Education and Research of Federal Republic of Germany, SiEGeN project, (BMBF, Grant No. 03X3546). The author would like to thank the people at CAU-Kiel (M. Manzoor, A. Malave, A. Petraru, C. Szillus, A. Priora, T. Metzing) and the people at Fraunhofer ISiT (J. Hagge, S. Käselau, M. Rickers, G. Nilson, F. Sörensen, Dr. L.-M. Buchmann, C. Eisermann, K. Reiter, B. Karsten, A. Ambrosius, D. Frank, B. Engel, K. Jung, D. Freidenberg, W. Prigandt, M. Witt, M. Claus).

References

1. S. Bhagat, H. Han, T.L. Alford, Thin Solid Films **515**, 1998–2002 (2006)
2. H. Kim, C. Cabral Jr., C. Lavoie, S.M. Rossnagel, J. Vac. Sci. Technol. B **20**(4), 1321–1326 (2002)
3. S.H. Ferguson, H.W. King, JCPDS Int. Cent. Diffr. Data **49**, 180–187 (2005)
4. R. Hübner, M. Hecker, N. Mattern, V. Hoffmann, K. Wetzig, C. Wenger, H.-J. Engelmann, C. Wenzel, E. Zschech, J.W. Bartha, Thin Solid Films **437**, 248–256 (2003)
5. Powder Diffraction Files (PCPDFWIN), JCPDS Int. Cent. Diffr. Data (1997), PDF#: 031051 (Ni), 231455(Ti_2N), 491440(TiW), 180893(Ni_3Ta), 251280(β-Ta), and 040788(α-Ta)
6. A. Leineweber, S. Kurz, M. Fonovic, Annual Report 2012, Max Planck Institute, Germany (2012)
7. N.S. Lawand, P.J. French, J.J. Briaire, J.H.M. Frijns, Procedia Engineering **47**, 726–729 (2012)

The Direct Observation of Grain Refinement Mechanism in Advanced Multicomponent γ-TiAl Based Structural Intermetallics Doped with Boron

A.V. Kartavykh, M.V. Gorshenkov and D.A. Podgorny

Abstract The synthesis of Ti–44Al–5Nb–2Cr–1.5Zr–0.4B–0.07La and Ti–44Al–5Nb–1Cr–1.5Zr–1B–0.17La (at.%) intermetallic alloys was performed from pure metals by the electron beam semi-continuous casting technique. Alloys are characterized by convoluted microstructure whose refinement is progressive with the boron content increase. The specimens were analyzed using JEOL JSM6610 scanning electron microscope (SEM) equipped with electron backscatter diffraction and energy dispersive X-ray spectroscopy (EDX) systems. The boron alloying causes precipitation of micro-dimensional complex (Ti,Nb)B borides within solidifying melt. These particles-inoculants promote the formation of finer microstructure. *Auger* spectrometry was applied for quantitative elemental analysis of borides, using PHI-680 device. It was proven that the origin of structure refinement consists in solid-phase seeding and germination of α_2-Ti_3Al-phase on (Ti,Nb)B ribbon-like colonies with subsequent growth of α_2-laths through the volumetrically prevailing γ-TiAl domains. We fixed several consecutive stages of microstructure refinement process. The description of these stages is supported with the dimensional, crystallographic and compositional characterization of microstructural constituents. The work provides a basis for new techniques of microstructure/mechanical properties engineering of studied materials.

1 Introduction

Lightweight multicomponent γ-TiAl-intermetallics are creep resistant to high temperatures (~ 1100 K), so being an advanced materials for turbine blades of aircraft engines and transportable gas-burning power-generation plants [1, 2]. Nb alloying can significantly improve high temperature properties of TiAl based alloys, whereas

A.V. Kartavykh (✉) · M.V. Gorshenkov · D.A. Podgorny
National University of Science and Technology "MISIS", Leninsky pr. 4, 119049 Moscow, Russia
e-mail: karta@korolev-net.ru

E.K. Polychroniadis et al. (eds.), *2nd International Multidisciplinary Microscopy and Microanalysis Congress*, Springer Proceedings in Physics 164,
DOI 10.1007/978-3-319-16919-4_23

minor additions of Cr and Zr enhance their oxidation resistance within turbine gas flow. Lanthanum addition was applied firstly by us for the internal gettering of undesired oxygen impurity [3]. Generally, finer microstructure provides better performance of an alloy that require the special two-phase (γ-TiAl + α_2-Ti_3Al) polycrystalline pattern creation in cast items. That is why the attention of researchers is presently paid to microstructure improvement by modifying doping of alloy with boron. The precise boron alloying causes precipitation of micro-dimensional complex Ti-based borides within solidifying melt. These particles-inoculants might act as seed centers promoting the formation of finer microstructure in an alloy when casting. Meanwhile, despite the practical application in cast TiAl, no clear-cut understanding of the grain refinement mechanism associated to boron additions is available today. The present paper reports the detailed characterization and analysis of microstructurally improved material, with the particular study of boride inoculants being formed within the alloy matrix, along with the direct observation of structure refining mechanism.

2 Materials and Methods

The synthesis of test ingots of alloys with Ti–44Al–5Nb–2Cr–1.5Zr–0.4B–0.07La and Ti–44Al–5Nb–1Cr–1.5Zr–1B–0.17La (at.%) nominal compositions has been performed by the electron beam semi-continuous casting technique in vacuum using the sticks of technically pure initial metals. Lanthanum hexaboride LaB_6 was used as the boron-contained additive. The alloys were cast in the shape of cylindrical billets of 160 mm in diameter and a mass of 15 kg. In the course of casting the different cooling rates were applied in the range from 1.3 to 3 K s^{-1}. More detailed description of applied cast technique could be found in paper [3].

The metallographic slices of alloys were analyzed using the *JEOL JSM6610* scanning electron microscope (SEM), equipped with *AZtecSynergy* combined electron backscatter diffraction (EBSD) and energy dispersive X-ray spectroscopy (EDX) system. The 1-μm diamond polishing of slices was done using *Buehler* equipment and consumables. No chemical etching was applied for the overall microstructure pattern reviewing by SEM. The deep etching was performed of slice surface with *Keller* etchant for the revealing of 3D-images of precipitated boride particles within the removed layer of alloy. We referred to the Inorganic Crystal Structure Database (ICSD), NIST and EBSD databases for the identification of revealed structural components. Since EDX has the fundamental restrictions regarding low sensitivity to the light elements, including the boron, the *Auger* spectrometry has been applied for the best recognizing and quantitative elemental analysis of borides, using the *PHI-680 Physical Electronics Auger Nanoprobe Analyser* device at 10 keV energy and 40 nm diameter of primary electron beam in 8×10^{-10} Torr vacuum. For the cleaning of analyzing surface from adsorbed atoms the ion gun was preliminarily applied with 1 keV Ar^+ beam ensuring on-depth sputtering rate of 5 nm min^{-1}. The sputtering performed until a steady state stabilization of signal level is achieved from all the detected elements in analyzing area. More detailed description of applied analytical techniques could be found in [4, 5].

3 Results and Discussion

Both cast alloys are characterized by convoluted microstructure whose refinement is progressive with the boron content increase. Engineering this structural type of TiAl-intermetallics is considered one of the most promised solutions for the material applications at extreme temperatures and mechanical loads [6]. As the best result the most fine-grained structure pattern of Ti–44Al–5Nb–1Cr–1.5Zr–1B–0.17La specimen is displayed in Fig. 1 along with its dimensional-statistic histogram. Due to boron doping, the material is synthesized/cast with the achieved mean grain size as small as 30 μm and rather perfect structural uniformity, where the standard deviation of grain diameter value (s) amounts 12.8 μm. X-ray diffraction and EDX analyses confirm the presence of intermetallic phases γ-TiAl and α_2-Ti_3Al as basic constituents with corresponding minor contents of alloying Nb, Zr and Cr species uniformly distributed in both phases. SEM observation suggests that α_2 is shaped in the form of structural laths, which mostly fill an inside area of primary solidified grains and pierce the volumetrically prevailing γ-TiAl matrix.

Besides, the randomly distributed ribbon-like curved boride colonies of several ten microns in length can also be distinguished in the γ-matrix (Fig. 2a). The application of different cooling rates in cast experiments allows tracing in details the structure refinement process that was terminated (frozen) respectively at different stages. Evidently, this process begins from the high-temperature solid-state germination (seeding) of α_2-Ti_3Al phase on boride ribbons in the (α + γ) domain of TiAl phase diagram (Fig. 2b), then proceeding through the stage of numerous

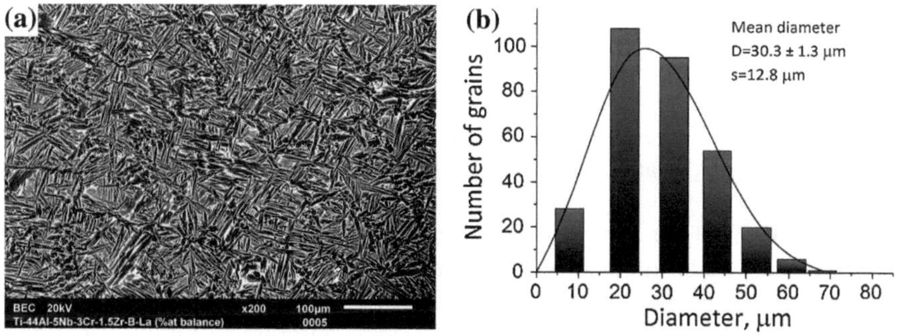

Fig. 1 **a** Microstructure (SEM micrograph) of Ti–44Al–5Nb–1Cr–1.5Zr–1B–0.17La (at.%) alloy; **b** the histogram of grains distribution by the diameter obtained by the processing of this image, and its approximation by the normal statistical distribution curve

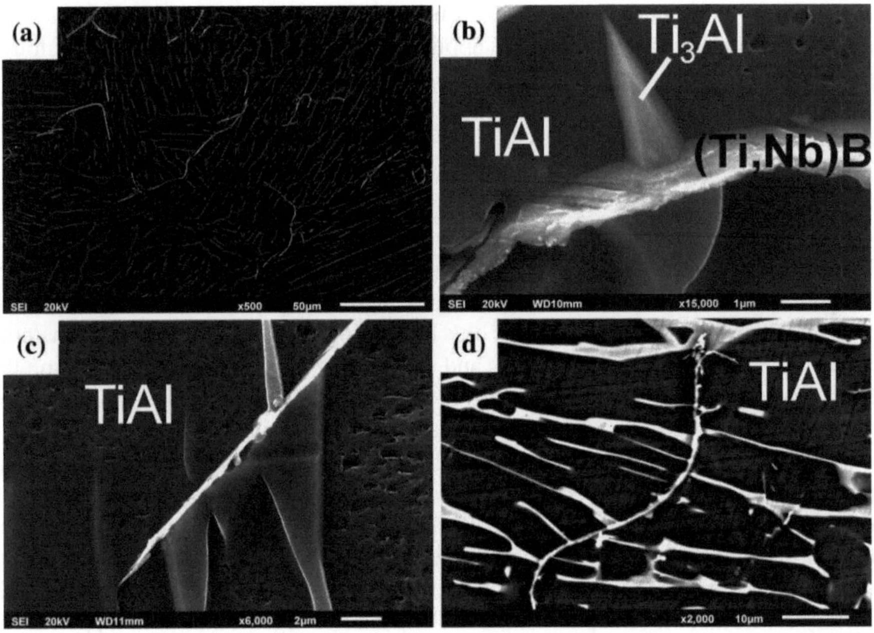

Fig. 2 Boride ribbons resolved in Ti–44Al–5Nb–2Cr–1.5Zr–0.4B–0.07La alloy by SEM (**a**) and three consecutive stages of structure refining mechanism development: the event of solid-state germination (seeding) of α_2-Ti_3Al phase on boride facet (**b**); morphological development of α_2-laths and their growth through the basic γ-TiAl phase domain (**c**); formation of new grain boundary by elongated curved boride surrounded by fine (γ-TiAl + α_2-Ti_3Al) ordered phase structure inside the delimited grains (**d**)

α_2-laths growth and morphological development (Fig. 2c). Finally it leads to new grain boundaries formation by the boride colonies, and ordered substructure creation of nearly parallel α_2-laths inside these grains (Fig. 2d) in alloy containing 0.4 at.% of boron. At the higher (1 at.%) boron content the formation of dense network of α_2-laths is characteristic (Fig. 3a). Boride and α_2 phases are hardly distinguishing in such SEM micrographs, but can be easily recognized by *Auger* or EDX elemental mapping techniques. In these patterns α_2-phase enriched by Ti and depleted in Al with respect to γ-background in correspondence with basic stoichiometric formula Ti_3Al (versus TiAl), and the boride needles reflect distinctly the boron content (Fig. 3b–d).

Examination of the typical boride ribbon revealed in the alloy matrix is displayed in Fig. 4a, b. *Auger* spectra suggests that the relative ratio of Ti, Nb and B contents within the nucleant particle nearly corresponds to 2:1:3 respectively, and the object contains no La at all. Therefore, most appropriate formula here could be expressed as (Ti,Nb)B, that reflects the new complex monoboride compound

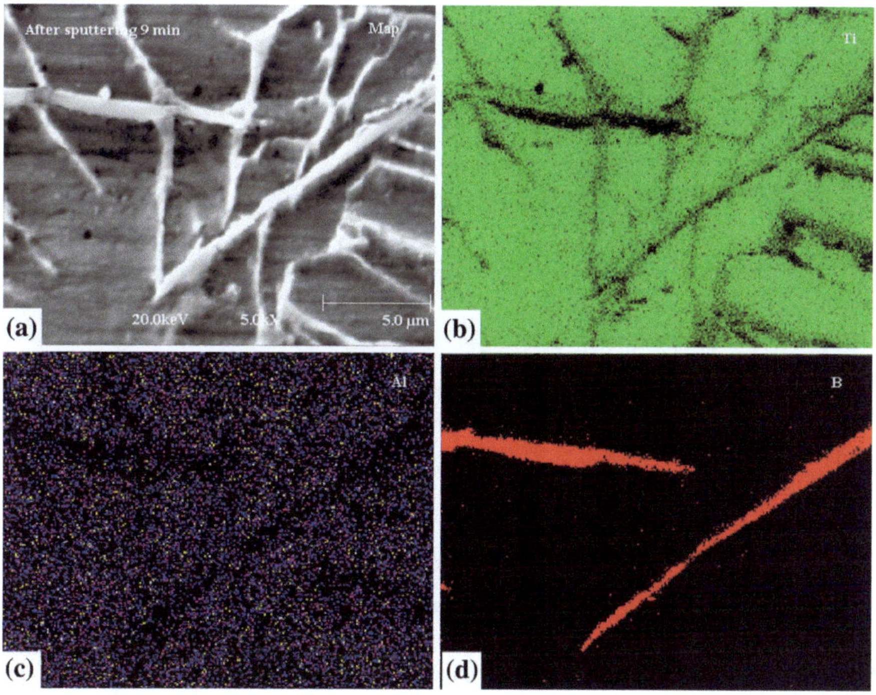

Fig. 3 Example of SEM image of α_2-network microstructure in Ti–44Al–5Nb–1Cr–1.5Zr–1B–0.17La intermetallic alloy in the vicinity of two boride ribbons (**a**) and the elemental maps of this area by *Auger* spectrometry: **b** titanium, **c** aluminum, **d** boron distributions

formation after complete dissolution of LaB_6 additive and consequent re-precipitation within a niobium-containing TiAl melt in the course of casting. Figure 4c, d represents the initial and indexed *Kikuchi* patterns, respectively, derived from this micro-precipitate by EBSD that match the object as TiB monoboride with orthorhombic structure *B27*, *Pearson* symbol *oP8*, spacegroup *Pnma*, no. 62 according to NIST database. Finally, with the account of *Auger* analysis data, the nature of re-precipitated ribbon may be evaluated as TiB monoboride with partial (by 1/3) substitution of Ti atoms by Nb at Ti-sublattice sites (nodes). It is known indeed that Nb is capable of substitutional solid solutions formation in TiB precipitated phase within wide range of Ti,Nb concentrations, since monoborides TiB and NbB have similar orthorhombic structure and practically near to be isomorphic [4, 5]. The substitution ratio (and, therefore, the crystallographic morphology and habit of re-precipitated particles) could be compositionally and thermodynamically dependent, realizing through the processing conditions of an alloy.

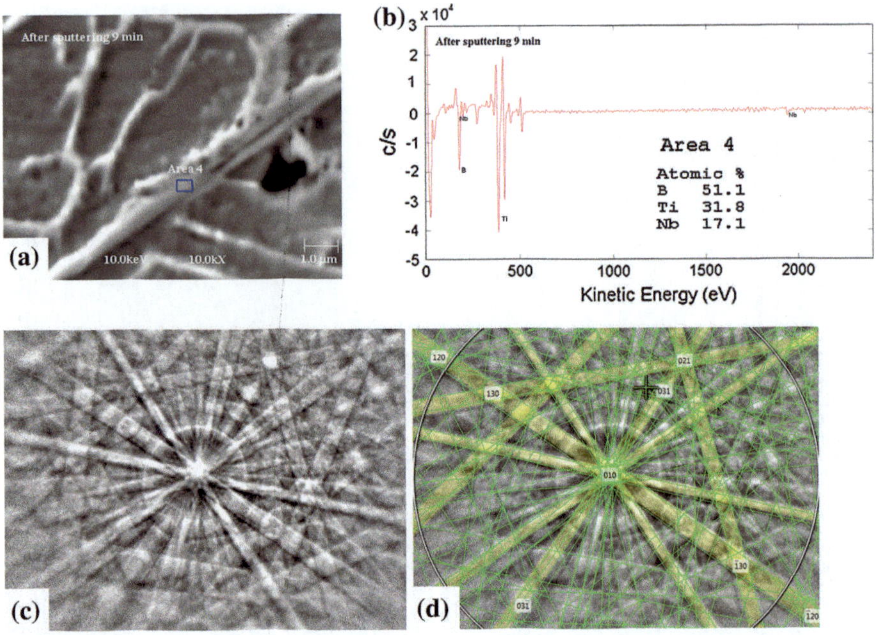

Fig. 4 **a** SEM image of typical boride precipitate in Ti–44Al–5Nb–1Cr–1.5Zr–1B–0.17La alloy with indication of sputtered analyzing area; **b** *Auger* spectrum derived from this area with determined atomic composition. The initial (**c**) and indexed (**d**) *Kikuchi* patterns derived from this monoboride by EBSD analysis using the *AZtec* software

4 Conclusions

1. The synthesis of alloy billets of Ti–44Al–5Nb–2Cr–1.5Zr–0.4B–0.07La and Ti–44Al–5Nb–1Cr–1.5Zr–1B–0.17La (at.%) compositions have been performed by the electron beam casting technique, applying LaB_6 compound as the boron alloying additive.
2. Both advanced cast alloys possess the convoluted microstructure whose refinement is progressive with the boron content increase. The best fine-grained specimen displays the mean structural grain size of 30 µm with 12.8 µm standard deviation.
3. It is stated that initial LaB_6 addition was dissolved in a melt and has been completely re-precipitated into complex boride elongated particles acting as randomly distributed seeds and promoting the phase microstructure refinement of intermetallics.
4. It is proved by SEM, EBSD and *Auger* spectral compositional nano-analysis that newly precipitated particles to be (Ti, Nb)B orthorhombic monoborides of *B27* structure (*Pearson* symbol *oP8*, space-group *Pnma*) grown as ribbon-like colonies where every third atom of Ti is substituted by Nb at Ti-sublattice sites.

5. It is firstly demonstrated that the origin of structure refining effect in as-cast intermetallics consists in high-temperature solid-phase germination of Ti_3Al-phase on (Ti, Nb)B ribbons with subsequent growth of α_2-laths through the volumetrically prevailing γ-TiAl domains.
6. Finally, we managed to trace and fix three consecutive stages of microstructure refining process development. The description of these stages is reported.

Acknowledgments Authors acknowledge the Russian Ministry of Education and Science grant 11.951.2014/K, and partly the Russian Foundation for Basic Research grant 13-03-00500.

References

1. A. Lasalmonie, Intermetallics **14**, 1123 (2006)
2. N.P. Lavery, D.J. Jarvis, D. Voss, Intermetallics **19**, 787 (2011)
3. A.V. Kartavykh, E.A. Asnis, N.V. Piskun et al., J. Alloys Comp. **588**, 122 (2014)
4. A.V. Kartavykh, V.V. Tcherdyntsev, M.V. Gorshenkov et al., Mater. Chem. Phys. **141**(2–3), 643 (2013)
5. A.V. Kartavykh, M.V. Gorshenkov, V.V. Tcherdyntsev, D.A. Podgorny, J. Alloys Comp. **586**, S153 (2014)
6. M.H. Loretto, Z. Wu, M.Q. Chu et al., Intermetallics **23**, 1 (2012)

Low Temperature Resistivity of the Rare Earth Diborides (Er, Ho, Tm)B_2

Jumat B. Kargin, C.R. Sebastian Haines, Matthew J. Coak, Cheng Liu, Alexander V. Matovnikov, Vladimir V. Novikov, Alexander N. Vasiliev and Siddharth S. Saxena

Abstract The discovery of superconductivity in MgB_2 sparked much interest in the diborides and led to further discoveries in this class of materials. Attention has focussed on the transitions metal diborides and the rare-earth diborides have remained sparsely investigated. Here we present electrical transport in polycrystalline samples of ErB_2, TmB_2 and HoB_2 as a function of temperature down to 0.1 K. Particularly interesting is the large difference in the temperature dependence of resistivity, above and below the clear magnetic ordering temperature of 13.8, 9 and 7.5 K for ErB_2, HoB_2 and TmB_2 respectively.

J.B. Kargin (✉) · C.R.S. Haines · M.J. Coak · C. Liu · S.S. Saxena
Cavendish Laboratory, University of Cambridge, Cambridge, UK
e-mail: kargin_db@enu.kz

C.R.S. Haines
e-mail: crsh2@cam.ac.uk

M.J. Coak
e-mail: mjc253@cam.ac.uk

C. Liu
e-mail: cl291@cam.ac.uk

S.S. Saxena
e-mail: sss21@cam.ac.uk

J.B. Kargin
Eurasian National University, Astana, Kazakhstan

A.V. Matovnikov · V.V. Novikov
Bryansk State University, Bryansk, Russia
e-mail: mator79@mail.ru

V.V. Novikov
e-mail: vvnovikov@mail.ru

A.N. Vasiliev
Moscow State University, Moscow, Russia
e-mail: vasil@mig.phys.msu.ru

E.K. Polychroniadis et al. (eds.), *2nd International Multidisciplinary Microscopy and Microanalysis Congress*, Springer Proceedings in Physics 164,
DOI 10.1007/978-3-319-16919-4_24

1 Introduction

The rare earth elements diborides (REE diborides) are hard, solid refractory materials that form part of a large family of diborides with the AlB_2 type hexagonal structures with space group *P6/mmm* [1]. There is one atom of metal and two atoms of boron per unit cell.

The REE diborides have a metallic conductivity. Some of them have magnetic phase transitions at low temperature. ErB_2, HoB_2 and TmB_2, as well as DyB_2 and TbB_2 become ferromagnetically ordered at low temperature [2–4], whereas YbB_2 is an antiferromagnet. Despite the study of these compounds [5–8], the electrical properties of many rare-earth diborides at low temperatures are unreported.

The aims of this work are to explore the temperature dependence of the electrical resistivity of RB_2 (R = Er, Tm and Ho) and to obtain additional information about the nature of the magnetic phase transitions REE diborides by measuring their temperature dependence of the resistivity.

2 Sample Preparation and Experimental Technique

The samples of ErB_2, HoB_2 and TmB_2 were kindly provided by Russian colleagues from the Moscow and Bryansk State Universities. Details of the synthesis can be found in [9].

The resistivity measurements of ErB_2, HoB_2 and TmB_2 were carried out using a HELIOX (Oxford Instruments) and an Adiabatic Demagnetisation Refrigerator (CMR).

The resistance of the samples was measured in a vacuum ($<1.33 \times 10^{-5}$ Pa) by a 4-contact method. The samples have the form of a cuboid. Contacts were made using DuPont Silver Epoxy 6838 and platinum wire (25 μm).

3 Results and Discussion

Experimental values of the resistivity $\rho(T)$ for REE diborides ErB_2, HoB_2 and TmB_2 are plotted against temperature in Figs. 1 and 2. The $\rho(T)$ of ErB_2, HoB_2 and TmB_2 shows a metal type of conductivity down to low temperature. Careful measurement of the geometry of the samples allowed us to calculate the resistivity from measured voltage to within a factor of 8 %. The uncertainty in the resistivity is still however dominated by the geometrical uncertainty.

The resistance of the sample falls monotonically with decreasing temperature. At 13.8 K there is an abrupt drop in the resistivity of the ErB_2 sample, which then displays markedly different temperature dependence down to a temperature of 0.1 K. We propose 13.8 K to be the onset of the ferromagnetic transition for the ErB_2 sample. In the HoB_2 and TmB_2 samples, similar changes in the temperature

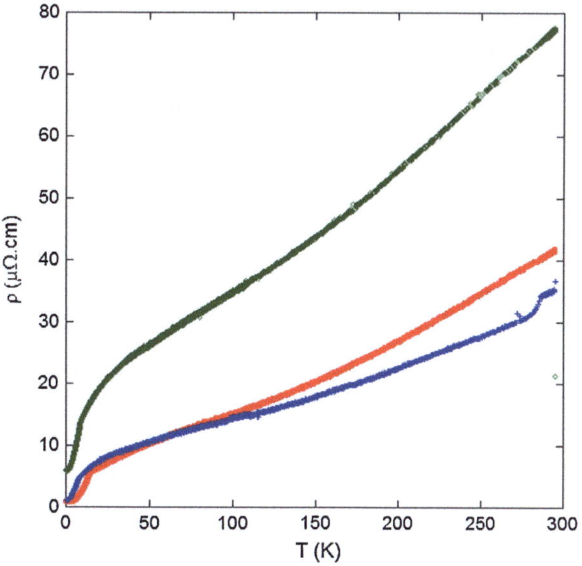

Fig. 1 Experimental data of the temperature dependence of the REE diborides resistivity (*red line* Erbium, *blue* Thulium, *green* Holmium)

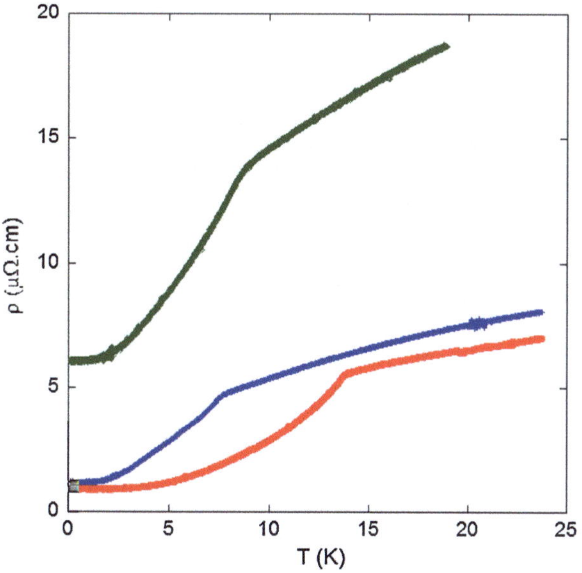

Fig. 2 Low temperature resistivity of the REE diborides (*red line* Erbium, *blue* Thulium, *green* Holmium)

dependence of their resistivity are observed at 9 and 7.5 K respectively, which we associate with the onset of the magnetic transition [2, 3].

The diborides do not exhibit superconductivity down to a temperature of 0.1 K. We have obtained residual resistivities of 7.5, 1.4 and 0.9 μΩ cm for ErB_2, HoB_2 and TmB_2 respectively.

Avilla et al. [5] measured the magnetic susceptibility and electrical resistivity of three samples of ytterbium diboride in the range 2–300 K. The resistivity was not calculated but the data is qualitatively similar to what we have seen in ErB_2, HoB_2 and TmB_2. The transition upon magnetic ordering is less pronounced in YbB_2. It is worth noting that YbB_2 is an antiferromagnet at low temperature unlike ErB_2, HoB_2 and TmB_2 which are all ferromagnets.

4 Conclusions

From our data we can say that we have seen evidence of the magnetic phase transitions in ErB_2, HoB_2 and TmB_2 at 14, 9 and 7.5 K respectively. The comparison of the resistivity data and the magnetic susceptibility data for REE diborides shows a small discrepancy. This discrepancy points to possibility of deviations from the standard theory of electron transport in metals at low temperatures and to the role of magnetism and dimensionality.

References

1. K.E. Spear, Chemical bonding in AlB2-type borides. J. Less-Common Met. **47**, 195–201 (1976)
2. A.V. Matovnikov, *Thermodynamic Properties of Rare-Earth Diborides*. PhD Thesis, Bryansk State University, Moscow, 2009
3. K.H.J. Buschow, *Boron and Refractory Borides,* ed. by Matcovich V.I. (Springer-Verlag, Berlin Heidelberg New York, 1977)
4. G. Will, V. Lehmann, K.H.J. Buschow, Magnetic properties and neutron diffraction of TbB2. Inst. Phys. Conf. Ser. **37**, 255–261 (1978)
5. M.A. Avila et al., Synthesis and properties of YbB2. J. Alloy. Compd. **358**, 6–64 (2003)
6. R.N. Castellano, Crystal growth of TmB2 and ErB2. Mater. Res. Bull. **7**, 261–266 (1972)
7. I.R. Shein, A.L. Ivanovskii, Effect of metal vacancies on the energy parameters of s-, p-, and d-metal diborides. Russ. J. Inorg. Chem. **52**(2), 238–241 (2007)
8. G. Will, W. Schafer, Neutron diffraction and the magnetic structures of some rare earth diborides and tetraborides. J. Less-Common Met. **67**, 31–39 (1979)
9. A.V. Matovnikov et al., Two-step syntheses of rare-earth diborides. Inorg. Mater. **45**(4), 366–368 (2009)

Influence of Mechanical Parameters on the Friction and Wear of Sliding Brass-Steel Couple

C. Boubechou, A. Bouchoucha and H. Zaidi

Abstract This paper presents an experimental study of the tribological behavior of a sliding contact pin-disc. This study concerns a brass-steel dry dynamic contact where friction and wear are studied in terms of mechanical parameters, without passage of electric current. To release this study, a tribometer, pin brass rubbing against a rotating steel disc is used in an atmospheric ambiance. The testing parameters are: the normal load (5–30 N) and sliding velocity (0.1–0.5 m/s). The duration of each test is 30 min. Experimental results show that these parameters have a significant effect on the tribological behavior of the studied couple. The discussion of results is based on observations, using a profilometer, worn surfaces and interfacial phenomena resulting from the sliding contact process.

1 Introduction

Contact between two bodies, they are static or dynamic, are the basis of all connections mechanisms. Often, contact between the elements displayed relative movement or tendency to this movement which produces micro-slips at the interface of contact [1]. The existence of these requires paying attention to problems

C. Boubechou (✉)
Faculté de Technologie, Département de Génie Mécanique, Université de 20 Août 1955, Skikda 21000, Algeria
e-mail: choubeila_boubechou@yahoo.fr

A. Bouchoucha
Laboratoire de Mécanique, Faculté des Sciences de la Technologie, Département de Mécanique Engineering, Université Constantine 1, Constantine 25000, Algeria
e-mail: bouchoucha_ali1@yahoo.fr

H. Zaidi
Laboratoire LMS (UMR-6610-CNRS), SP2MI, Téléport 2, Boulevard Marie et Pierre Curie, Université of Poitiers, BP 30179, Futuroscope Chasseneuil Cedex Poitiers 86962, France

E.K. Polychroniadis et al. (eds.), *2nd International Multidisciplinary Microscopy and Microanalysis Congress*, Springer Proceedings in Physics 164,
DOI 10.1007/978-3-319-16919-4_25

of friction forces induced on the surface and which are in addition to the normal forces between the contact elements. Therefore, the relative movement to the contact interface may be an additional source of friction that plays an important role in the damage of the surfaces. The friction causes the deterioration of opposing surfaces, it can take different forms: plastic deformation, wear, contact fatigue, etc. [2].

The aim of this paper is to study the influence of the parameters: load, time parameter and sliding speed on the tribological behavior of the sliding brass-steel couple without electrical current.

2 Experimental Device

The experimental device is a pin-on-disc tribometer. Tests are carried out in an ambient air (Fig. 1). The pin is made of brass and the disc of steel (Tables 1, 2 and 3). The disc is driven in a rotational movement of constant velocity with diameter of 50 and 12 mm of thickness. It has a lame hole in the center. The normal load is vertically applied with a weight P. The pin has a cylindrical shape of length 20 mm and a diameter of 8 mm; has a flat allows to fix it in a hole with a screw on an arm load of aluminum. Friction coefficient μ is deduced from measurement of the tangential force F induced on the arm by the rotating wheel through the pin ($\mu = F/P$). The wear W is determined by weighing the specimen before and after each test using a balance with a precision of ±0.1 mg. To ensure the same experimental conditions for each test, the disc is polished mechanically with a 1200 grain paper.

3 Materials

Tables 1, 2 and 3.

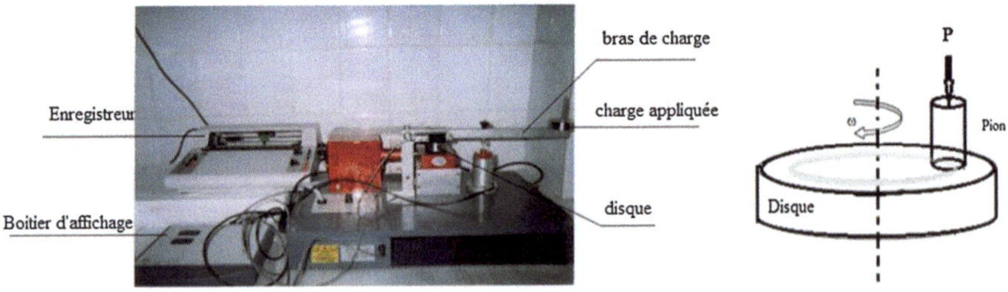

Fig. 1 Tribometer pin-disc

Table 1 Chemical composition of the pin and the disc

	C (%)	Si (%)	Mn (%)	P (%)	S (%)	Cr (%)	Mo (%)	Cu	Zn	Pb
42CrMo4	0.38–0.45	0.015–0.4	0.5–0.8	0.035	0.035	0.90–1.2	0.15–0.3			
CuZn40Pb3	–	–	–	–	–	–	–	57	40	3

Table 2 Mechanical characteristics of the pin and the disc

	E (N/mm^2)	R (N/mm^2)	R_m (N/mm^2)	A (%)	HB
42CrMo4	210,000	650	605	12	270–330
CuZn40Pb3	96,000	180	460	15	120

Table 3 Physical characteristics of the pin and the disc

	ρ (Kg/m^3)	C_p (J/Kg °C)	λ (W/m °C)	γ (μ Ω m)
CuZn40Pb3	8.5×10^3	377	121	0.062
42CrMo4	7.850×10^3	480	44	24×10^{-2}

4 Experimental Results

Figure 2 shows the evolution of friction coefficient versus time the brass and steel couple. On this curve there are two distinct areas:

(a) First area (t = 0–15 min): it is a transient phase (lapping), in which the coefficient of friction varies between two extreme values.
(b) The second area (t = 15–30 min): it characterizes a stationary state, wherein the state of equilibrium is determined by stability of the operating conditions at the interface. For the second area, the coefficient of friction is stabilized at the value $\mu = 0.15$.

Figure 3 illustrates the variation of the friction coefficient (mean) according μ of the normal load applied to the steel-brass torque. In reviewing this figure, the curve shows a linear dependency of the load, which leads to a substantially constant coefficient of friction ($\mu \sim 0.17$).

The curve 4 shows the variation of μ with the sliding speed. On this graph, we can see two different branches:

- 0.1 V ≤ ≤ 0.3 m/s: the first part is characterized by relative stability of the μ limit value 0.2.
- 0.3 V ≤ ≤ 0.5 m/s: in the second part, μ decreases significantly as the speed increases, reaching a maximum of about 0, 15 value.

Figure 4 shows that the sliding speed does not have a significant effect on the coefficient of friction.

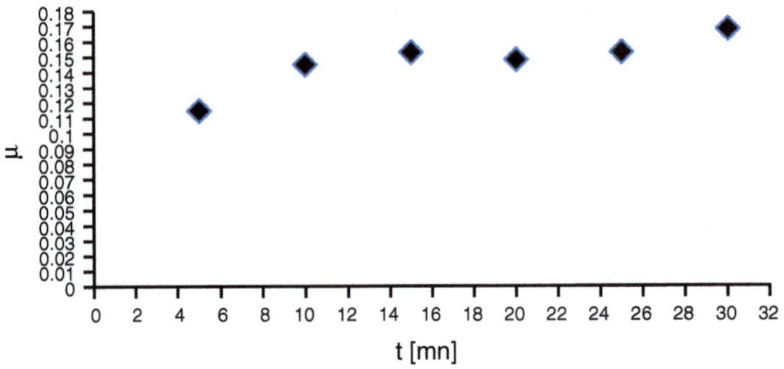

Fig. 2 Variation of friction coefficient versus time (P = 20 N and V = 0.5 m/s)

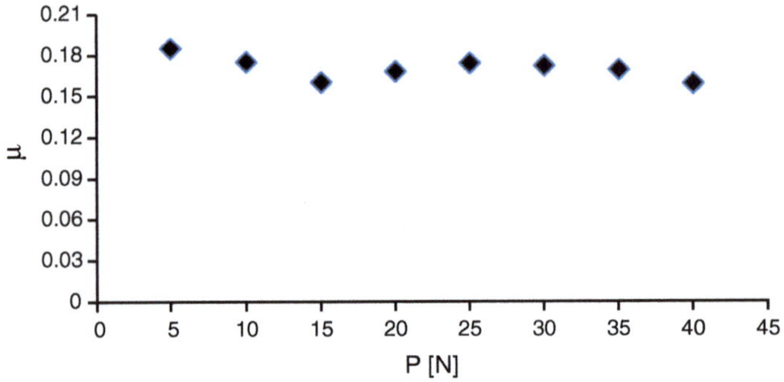

Fig. 3 Variation of friction coefficient versus normal load (t = 30 mn and V = 0.5 m/s)

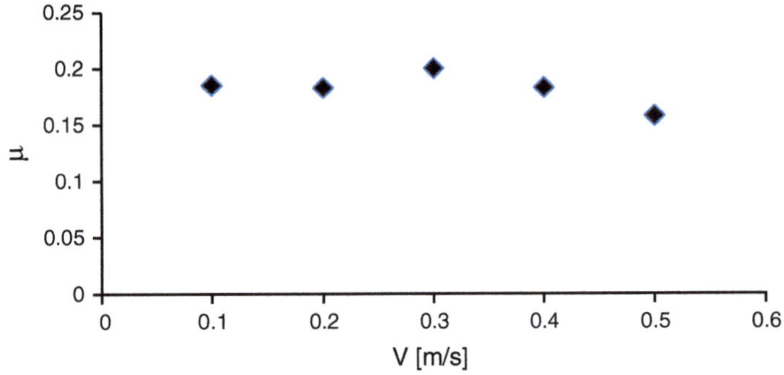

Fig. 4 Variation of friction coefficient versus sliding speed (P = 20 N and t = 30mn)

5 Analysis and Interpretation of Results

5.1 *Effect of the Time Parameter to Friction and Wear*

During the 30 min, the contact between the elements of the pair is of a metallic nature mutual transfers interface are also metal [3]. There is very little debris in the ignition and the coefficient of friction is low (μ = 0.11). With time the debris are formed, particularly of hard copper particles are pulled and they play a role in the abrasive-steel brass torque, which increases the friction coefficient. The sudden increase in the coefficient of friction during the first moments of friction (Figs. 1 and 4) can be explained by interactions between the asperities of opposing surfaces. Indeed, as the dynamic contact continues, the adaptation of surfaces occurs and the mechanism of formation and rupture junctions stabilizes balance operating conditions at the interface (late break) [4]. This increase is explained by the increase in the real contact surface deformation and creep. This surface shape generally elliptical deforms and elongates in the direction of sliding. As against, in the second zone, the coefficient of friction is stable. This stability is due to the accommodation of the friction interface. Asperities are sheared and there is a significant production of fine particles from both surfaces [5]. The production of particles and increase the stability of the temperature induces a friction coefficient due to changes in structure and composition as well as the phenomenon of surface oxidation process surfaces and the mutual transfer of the particles plays an important role the friction and wear behavior of this couple [6].

5.2 *Effect of Normal Load on the Friction and Wear*

At low loads, the surfaces of materials antagonists touch only by the highest asperities. The surface of contact increases as the load P increases under the combined action of the normal load and the shear stress action asperities entangle to form the actual contact surface [7]. After lapping the surfaces become smooth and fit, the friction coefficient decreases and stabilizes when the operating conditions are stable at the interface [8] (Fig. 2).

The wear of the pin is low due to the limited number of points of contact between the surfaces and the important role of adsorbed as a lubricating element [9] (Fig. 5) layer. When increasing the load, this translates into an increase of the actual contact area and the temperature at the interface, resulting in different effects and especially facilitates the formation and removal of oxide [10]. Figure 5 shows that the sliding speed does not have a significant effect on the coefficient of friction.

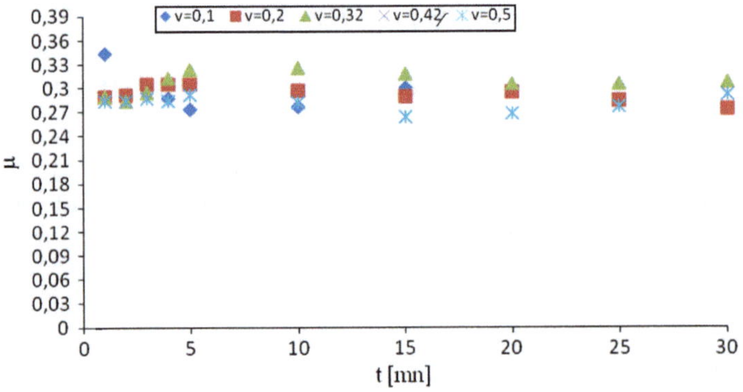

Fig. 5 Variation of friction coefficient versus time for different values of parameter speed (P = 20 N and I = 0 A)

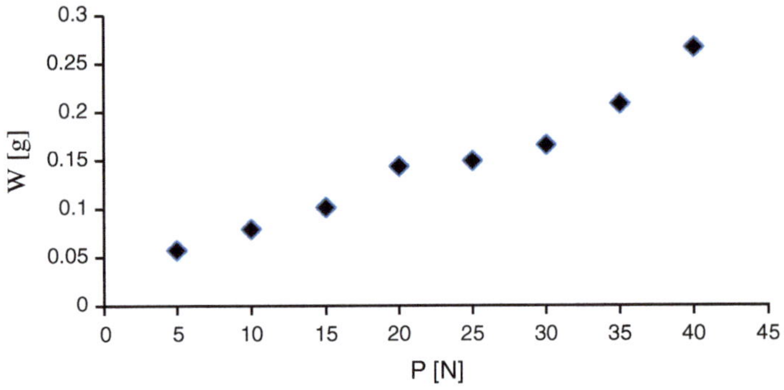

Fig. 6 Variation of the wear versus normal load (t = 30 mn and V = 0.5 m/s)

5.3 Effect of Speed on Friction and Wear

The sliding velocity changes the duration of surface exposure to the surrounding environment, but it affects the amplitude of the local temperature rise in the contact [11].

A low speed (0.1 m/s) where the exposure time is large and the temperature rise is low, the friction coefficient is about a moderate value of the order of 0.3 (Fig. 3).

In the range of 0.1–0.2 m/s, the coefficient of friction μ and the wear pins W (Figs. 3 and 6) are raised to the top because the metal junctions have the time needed sections to increase their creep and actual contact surface widens in the direction of movement [12]. The decrease in the coefficient of friction and wear is attributed to the decrease in rise time of the junctions, and therefore the number of metal contacts is reduced by forming an oxide layer on the friction face of the pin which reduces friction and protects the surface against damage by ensuring good wear resistance [12] (Fig. 7). The curve 6 shows the variation of the wear as a

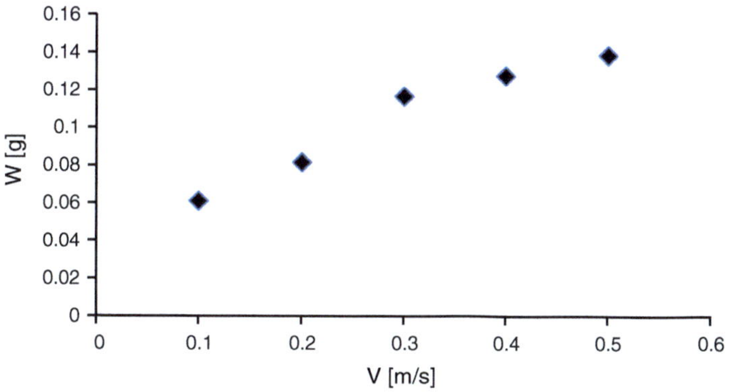

Fig. 7 Variation of the wear versus sliding speed (P = 20 N and t = 30mn)

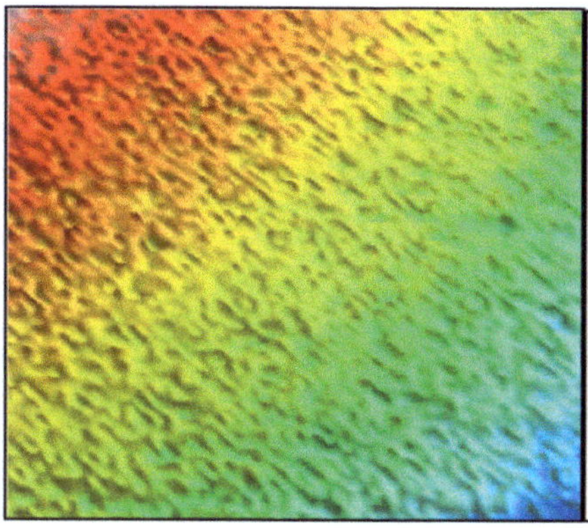

ISO 25178			
Paramètres de hauteur			
Sq	3.23826	µm	Hauteur quadratique moyenne de la surface
Ssk	-0.0959396		Facteur d'asymétrie de la surface
Sku	2.39155		Facteur d'aplatissement de la surface
Sp	7.57164	µm	Hauteur maximale des saillies de la surface
Sv	8.78933	µm	Hauteur maximale des creux de la surface
Sz	16.3610	µm	Hauteur totale de la surface
Sa	2.66170	µm	Hauteur arithmétique moyenne de la surface
Paramètres fonctionnels (généraux)			
Smr	0.834129	%	Taux de portance surfacique
Smc	4.32879	µm	Hauteur de taux de portance surfacique

Fig. 8 Optical image obtained by profilometer a pin brass before the test

function of the normal load. It is found that the wear is almost proportional to the normal force. For Fig. 7, the wear changes linearly according to the sliding speed.

In the second zone or the velocity V is greater than 0.2 m/s, the temperature at the interface is higher, thus increasing the speed of the oxide layer [3] therefore oxidation rate. The oxide film is broken, an almost constant friction coefficient and a higher wear is observed. The worn face of the pin has colored areas, materializing the heat affected region. The transfer of metal to the disk increases, the friction surface is deformed with furrows plowed in the direction of movement of an abrasive wear characteristic (Figs. 9 and 10).

The examination of the worn surface of the brass pin optical profilometer that shows the sliding speed has a significant effect on the surface quality (roughness) of the pin. This analysis also showed that the maximum height of the protrusions and recesses decreases with the linear sliding velocity (Figs. 8, 9 and 10).

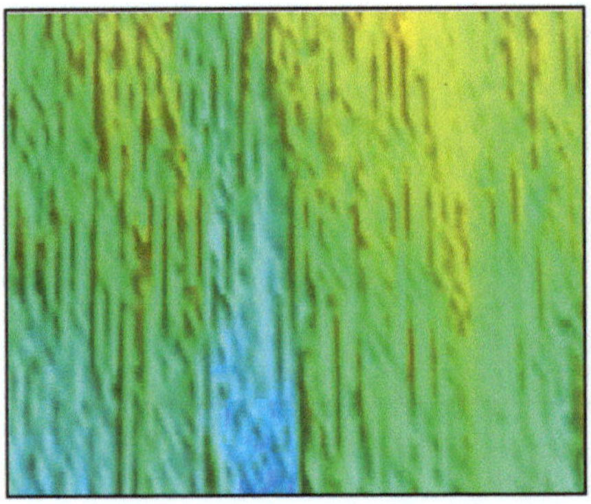

ISO 25178			
Paramètres de hauteur			
Sq	5.51124	µm	Hauteur quadratique moyenne de la surface
Ssk	0.370407		Facteur d'asymétrie de la surface
Sku	2.71917		Facteur d'aplatissement de la surface
Sp	21.5247	µm	Hauteur maximale des saillies de la surface
Sv	13.7974	µm	Hauteur maximale des creux de la surface
Sz	35.3221	µm	Hauteur totale de la surface
Sa	4.51730	µm	Hauteur arithmétique moyenne de la surface
Paramètres fonctionnels (généraux)			
Smr	0.0294291	%	Taux de portance surfacique
Smc	7.34327	µm	Hauteur de taux de portance surfacique

Fig. 9 Optical image obtained by profilometer a pin brass (P = 20 N and V = 0.4 m/s)

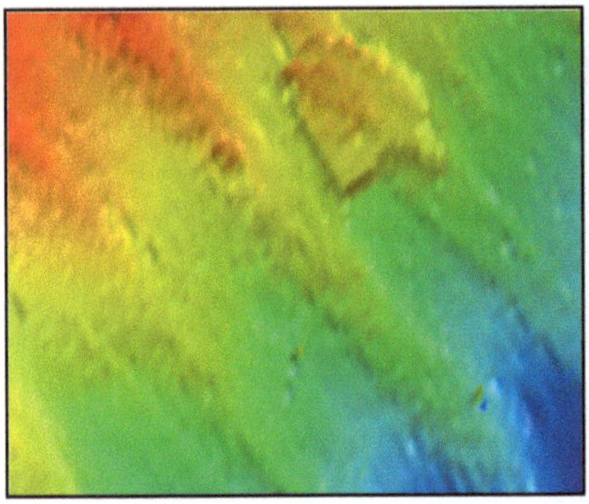

ISO 25178			
Paramètres de hauteur			
Sq	13.9975	µm	Hauteur quadratique moyenne de la surface
Ssk	-0.0385307		Facteur d'asymétrie de la surface
Sku	1.95629		Facteur d'aplatissement de la surface
Sp	31.0373	µm	Hauteur maximale des saillies de la surface
Sv	32.9783	µm	Hauteur maximale des creux de la surface
Sz	64.0156	µm	Hauteur totale de la surface
Sa	12.0174	µm	Hauteur arithmétique moyenne de la surface
Paramètres fonctionnels (généraux)			
Smr	0.00982029	%	Taux de portance surfacique
Smc	18.7940	µm	Hauteur de taux de portance surfacique

Fig. 10 Optical image obtained by profilometer a pin brass (P = 20 N and V = 0.5 m/s)

6 Conclusions

It appears from this experimental study that the mechanical parameters play a significant role in the tribological behavior of brass-steel sliding contact, in fact:

- The friction coefficient has a linear dependency with the sliding speed.
- Wear increases proportionally with the normal load.
- Wear increases linearly with the sliding speed.

References

1. A. Swartz, *Glossary of Terms and Definitions in the Fields of Friction* (Organization for Economic Coperation and development (OCDE), 1969)
2. A. Bouchoucha et al., Influence of the electrical sliding speed on friction and wear processes in electrical contact copper–stainless steel. Appl. Surf. Sci. **223**, 330–342 (2004)

3. A. Bouchoucha et al., Influence of electric fields on the tribological behavior of electrodynamical copper/steel contacts. Wear **203–204**, 434–441 (1997)
4. E. Hounkponou et al., Tribological behavior of graphite/graphite and graphite/copper couples in sliding electrical contact: influence on the contact electric field on the surface passivation. Appl. Surf. Sci. **70/71**, 176–179 (1993)
5. H. Zahouani et al., Effect of roughness scale on contact stiffness between solids. Wear **04**, 067 (2008)
6. S. Jacobson, S. Hogmark, Surface modifications in tribological contacts. Wear **04**, 035S (2008)
7. D.H. Hwang et al., Influence of wear particle interaction in the sliding interface on friction of metals. Wear **225–229**, 427–439 (1999)
8. M.O.A. Mokhtar et al., A theoretical and experimental study on the dynamics of sliding bodies with dry conformal contacts. Wear **218**, 172–178 (1998)
9. J.B. Singh et al., Dry sliding of Cu–15 wt%Ni–8 wt% Sn bronze: wear behavior and microstructures. Wear **263**, 830–841 (2007)
10. D.M. Rawson, T.F.J Quinn, Frictional heating and the oxidational theory of wear. J. Phys. **13** (2), 19–209 (1980)
11. M. Amirat, H. Zaïdi J. Frêne, Effet de l'environnement sur le processus d'adhérence des particules d'usure sur les surfaces du contact magnétisé du couple laiton-acier. 18ème Congrès Français de Mécanique Grenoble (2007)
12. H. Werkstofftechnik, Wear and wear mechanisms: tribology. Mater. Sci. Eng. Lotharstr 1, 47057 Duisburg, Germany (2008)

Measuring the Degree of Sensitization (DOS) Using an Electrochemical Technique

Mokhtar B. Abuzriba and Salem M. Musa

Abstract Sensitization can be simply defined as the susceptibility of an alloy, specifically austenitic stainless steel, to corrosion at grain boundaries. A detailed study on types 304 stainless steel has been carried out to correlate the degree of sensitization measured by electrochemical potentiokinetic reactivation test (EPR) with the susceptibility to intergranular corrosion. In this study four different heat treatments were given to alloys, i.e., solution annealed (SA) at 1020 °C for 1 h, then quenched in water; also then heat-treated in air at 620 °C for 15, 30, and 60 min. The electrolyte for the EPR tests was 1.0 N H_2SO_4 solution containing 0.01 M KSCN. Potentiodynamic curves from passive to active regions in deaerated solution at room temperature were obtained at a scan rate of 1.67 mV/s (6 V/h), after the passivation at 200 mV versus (SCE) for 2 min then the polarization was conducted. The criterion used to distinguish between sensitized and non-sensitized specimens is the activation charge Pa, the peak current density, Ph, in the active state, and Flade potential Ef at which the active curve breaks upward. The results indicated that the longer the sensitization time the higher the activation charge (Pa), and the higher the peak current density in the active state (Ph). The results indicated that, the EPR is more sensitive than the chemical method for measuring the degree of sensitization.

1 Introduction

Sensitization can be simply defined as the susceptibility of an alloy, specifically austenitic stainless steel, to corrosion at grain boundaries. Grain boundaries are disordered misfit regions separating grains of different crystallographic orientation.

M.B. Abuzriba (✉) · S.M. Musa
Materials & Metallurgical Engineering Department, University of Tripoli, P.O. Box 13304, Tripoli, Libya
e-mail: dr.mokhtarphd@yahoo.com

S.M. Musa
e-mail: sm_musa@yahoo.com

E.K. Polychroniadis et al. (eds.), *2nd International Multidisciplinary Microscopy and Microanalysis Congress*, Springer Proceedings in Physics 164,
DOI 10.1007/978-3-319-16919-4_26

Intergranular corrosion of austenitic stainless steels is thought to be caused by one of three things: impurities at the grain boundaries, enrichment of one of the alloying elements, or the depletion of one of these elements in the grain boundary area. However, the most universally accepted theory for intergranular corrosion is based on the depletion of chromium in the grain boundary areas [1].

The depletion of Chromium at the grain boundaries results when the temperature is raised into the 400–850 °C range or during slow cooling from the austenitic temperature range, and is caused by the formation of $Cr_{23}C_6$. This leaves the Chromium content near the grain boundaries lowered to the point where the depleted area approaches ordinary steel rather than stainless steel.

Historically intergranular corrosion was recognized and studied using acid immersion tests. Acid tests exist which have been standardized by the ASTM. The Copper Copper sulfate–sulfuric acid test (ASTM A393-63), commonly known as the strauss test.

Recent research on electrochemical techniques for studying the susceptibility of stainless steels to intergranular attack indicate that the corrosion is primarily under anodic control. That is, attack is determined by the availability of anodic sites at the grain boundaries. It has been found that the anodic polarization curve of a sensitized stainless steel is different than the curve for a non-sensitized stainless steel [2, 3].

Trends in the electrochemical potentiodynamic reactivation method (EPR) has been investigated by a number of researchers [4–8]. They studied the measurement of the area that defined the active-passive region as a basis for assessing the degree of sensitization in stainless steel. Their study was prompted by the desirability of having a rapid, non-destructive field test for determining the degree of sensitization. They concluded that the EPR test remains the most sensitive electrochemical testing technique in detecting the degree of sensitization.

2 Materials and Methods

Degree of sensitization (DOS) The DOS studies were conducted on type 304 stainless steel using an electrochemical technique called the Electrochemical Potentiokinetic Reactivation method (EPR). The chemical composition of this alloy is given in Table 1.

In this study four different heat treatments were given to the alloys, i.e. solution annealed (SA) at 1020 °C for 1 h, then quenched in water; also then heat-treated in air at 620 °C for 15, 30, and 60 min. The electrolyte for the EPR tests was 1.0 NH_2SO_4 solution containing 0.01 M KSCN. The specimens were polished through

Table 1 Chemical composition of 304 stainless steel studied (weight percents)

C	Mn	Si	Cr	Ni	P	S	Fe
0.07	2.00	1.00	18.5	8.5	0.04	0.03	Bal.

Table 2 EPR test condition

Electrolyte	1.0 NH_2SO_4 + 0.01 M KSCN
Surface finish	1 μm alumina slurry
Temperature	25 °C
Deaeration	High purity N_2
Deaeration time	20 min
Deaeration rate	250 cm^3/min
Conditioning time	5 min
Passivation potential	200 mV for 2 min
Reactivation scan rate	1.67 mV/s
Polarization system	EG and G model 350A
Auxiliary electrodes (2)	C
Reference electrode	Standard calomel

1 alumina slurry. Potentiodynamic curves from passive to active regions in deaerated solution at room temperature were obtained at a scan rate of 1.67 mV/s (6 V/h), after the passivation at 200 mV (SCE) for 2 min. The polarization was conducted under the conditions given in Table 2. The criterion used to distinguish between sensitized and non-sensitized specimens is the activation charge Pa (given by the integrated area below the peak of the reactivation curve), the peak current density Ph, in the active state, and the Flade potential Ef at which the active curve breaks upward. Sensitized steels show higher (Pa) and (Ph) values and more positive (Ef) values than non-sensitized steels.

Metallography The specimens that were used for the Electrochemical Potentiokinetic Reactivation tests were electrolytically polished and etched using 60 % H_3PO_4 + 40 % H_2SO_4 and 10 % ammonium persulfate respectively. The microstructure was observed using light microscopy at magnifications of 100× and 1000×.

ASTM A262-Practice E The degree of sensitization was observed in this investigation by the use of the Copper-Copper sulfate-sulfate acid test. The test specimens were exposed to the boiling test solution for 24 h. Then the damage was assessed by bending the specimens through 180 and examining the outside surface for accentuated intergranular penetrations. The specimens sensitized for 30 and 60 min at 620 °C exhibited fissuring when bent, which is evidence of intergranular attack.

3 Results and Discussions

The susceptibility to intergranular corrosion was evaluated electrochemically by developing potential sweep from passive to the active region (reactivation) in 1.0 NH_2SO_4 + 0.01 M KSCN at 25 °C as the test solution.

Table 3 Result of Potentiokinetic Reactivation Measurements for 304 Stainless steels

Heat treatment	Pa (C/cm^2)	Ph (NA/cm^2)	Ef (V)	Ecorr (V)
Solution quenched (SQ)	0	9.95 × 102	−0.190	−0.314
SQ + 15 min (620 °C)	0.022	3.49 × 105	−0.100	−0.448
SQ + 30 min (620 °C)	0.244	3.61 × 106	−0.70	−0.459
SQ + 60 min (620 °C)	0.392	5.72 × 106	−0.050	−0.461

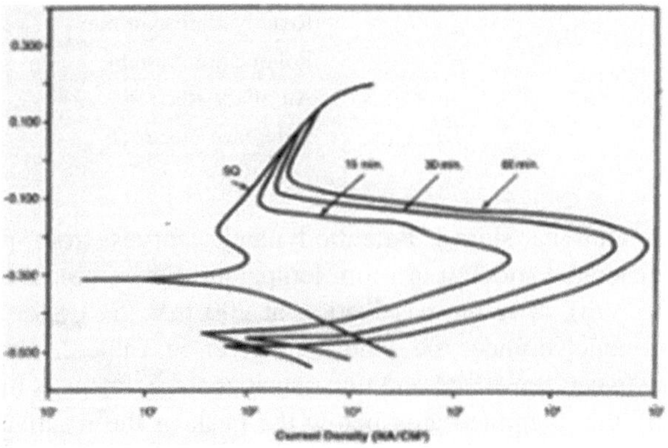

Fig. 1 Potentiokinitic reactivation curve, for solution quench (SQ) sample, sensitized at 620 °C and tested in 1.0 NH_2SO_4 + 0.01 M KSCN at 25 °C

The results of the Electrochemical potentiokinetic reactivation test (EPR) have been summarized in Table 3 and shown in Fig. 1. It can be seen that the longer the sensitization time the higher the activation charge (Pa) the peak current density in the active state (Ph) and that the activation charge appear to be more sensitive. It is obvious that considerable difference exists between the solution quenched (SQ) and the sensitized specimens. For Type 304 stainless steel, the sensitization is reflected in the variable degree of grain boundary attack (pitting) after etching the specimens using 10 % ammonium persulfate as shown in their microstructures, Figs. 2, 3, 4 and 5. The pitting attack along the grain boundaries, which is probably connected to the carbide structure of this material, and the amount of carbide precipitation, supports the idea that attack is associated with the Cr-depleted area. The results of ASTM A262-E indicate that specimens sensitized for 30 and 60 min exhibited fissuring when bent through 180, which is evidence of intergranular attack. These results agree quite well with the EPR measurements. In addition, the A262-E results for the 15 min sensitized specimens indicate doubtful test results, which is in contrast to that determined by EPR testing.

Precipitation of chromium carbides was suspected in specimen sensitized for 15 min. Observation under a TEM did not, however, reveal precipitates of chromium carbides but chromium was seen diffusing toward the grain boundaries as compared to a solution quenched specimen of the same alloy. This seems to reaffirm that the EPR is a very sensitive technique for measuring the degree of sensitization (susceptibility to intergranular corrosion).

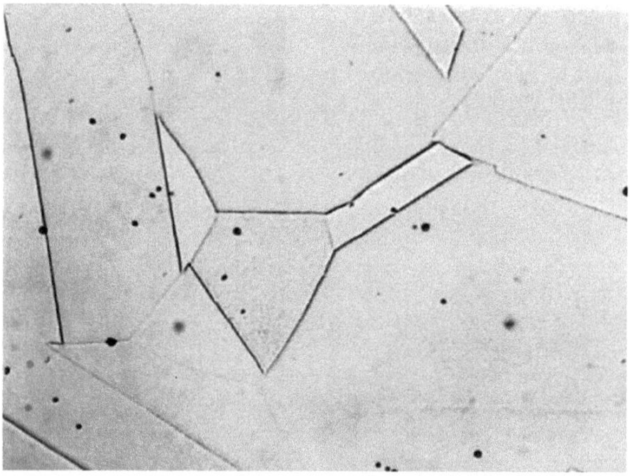

Fig. 2 Microstructure of solution heat treated sample for 1 h, then water quenched (SQ) 1000×

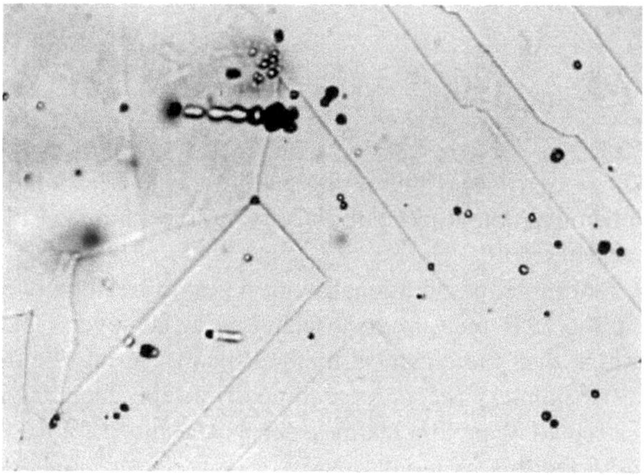

Fig. 3 Microstructure of solution heat treated sample for 1 h, then sensitized for 15 min at 620 °C 1000×

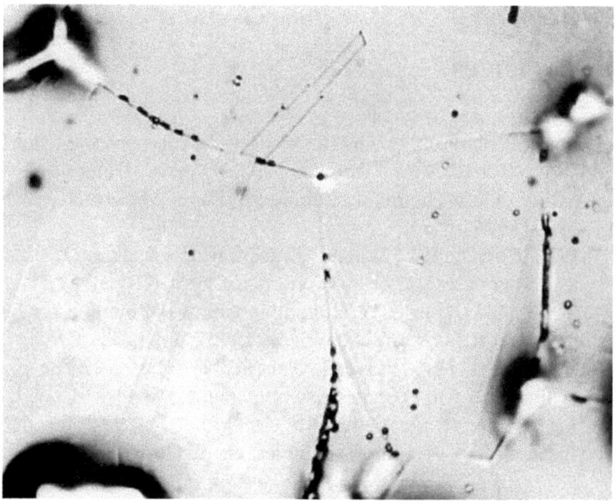

Fig. 4 Microstructure of solution heat treated sample for 1 h, then sensitized for 30 min at 620 °C 1000×

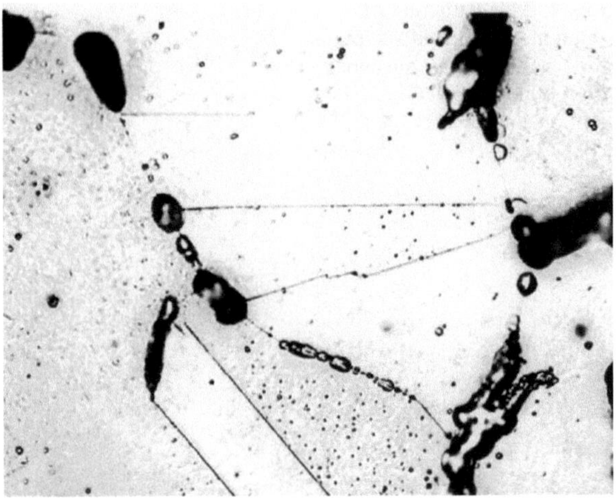

Fig. 5 Microstructure of solution heat treated sample for 1 h, then sensitized for 1 h at 620 °C 1000×

4 Conclusions

The objective of this investigation was to determine the feasibility of the Electrochemical Potentiokinetic Reactivation technique (EPR) for measuring the degree of sensitization.

Admittedly, additional work needs to be done before a high level of confidence in the EPR measurement technique is achieved. The EPR technique is more sensitive than the chemical method for measuring the degree of sensitization in Type 304 Stainless Steel. As a comparison, the chemical method would probably have accepted Type 304 stainless steels sensitized for 15 min, while the EPR technique rejected this material.

References

1. M.G. Tontana, N.D. Greene, *Corrosion Engineering* (McGraw-Hill, Inc. New York, 1979)
2. W.L. Clark, V.M. Romero, J.C. Danko, Detection of sensitization in stainless steel using electrochemical technique. Paper presented at NACE corrosion 77. Report No 180 (1979)
3. W.L. Cowan, R.L. Walker, Detection of sensitization in stainless steel using electrochemical technique. General electric report for US—NRC. CEAP Z1382 (1976)
4. V. Cihal, J. Klimsza, Elaboration of the EPR method for high alloyed steel. In *Proceedings of the EUROCORR*, 96 Nice XP5, (1996), pp. 1–4
5. V. Cihal, J. Hubdckove, J. Kubelka, K. Mazanec, The potential polarization method for the evaluation of martenzite-austenite stainless steels. Mater. Chem. Phys. **11**, 279 (1984)
6. J.R. Scully, R.Q. Kelly, An electrochemical test for detecting the intergranular corrosion susceptibility of a duplex stainless steel. Corrosion (NACE) **42**, 537 (1986)

7. G.H. Wagner, The EPR technique to evaluate special questions on the corrosion resistance of high alloyed materials. Werkstoffe u. Korrosion **52**, 15 (2001)
8. V. Cihal, S. Lasck, M. Blahetova, E. Kalabisova, Z. Krhutova, Trends in the electrochemical polarization potentiodynamic reactivation method—EPR. Chem. Biochem. Eng. Q **21**(1), 47–54 (2007)

Substitution for Chromium and Nickel in Austenitic Stainless Steels

Mokhtar B. Abuzriba and Salem M. Musa

Abstract In this investigation three alloy compositions of Fe–Al–Mn system were prepared for corrosion testing using an arc furnace with a protective atmosphere of argon, and using commercial iron, electrolytic manganese, pure aluminum, silicon and carbon. The alloys were melted into 75 g bar—shaped ingots. The anodic polarization, cyclic polarization and Tafel extrapolation techniques were applied to study and evaluate the corrosion resistance of austenitic Fe–Al–Mn alloy steels in a sulfuric acid adjusted to a pH of 2.4 and in 3.56 wt% NaCl. Though the mechanical properties and the oxidation resistance of these alloys have been reported to be comparable or superior to 18 % chromium-8 % Nickel austenitic stainless steels, the corrosion resistance was found to be poor and certainly not comparable to those of 18 % Chromium-8 % Nickel stainless steels.

1 Introduction

The term "Stainless Steel" refers to a ferritic iron-base alloy system with at least 10–12 % Cr or to an austenite iron-base alloy system with nickel or manganese and at least 17–18 % chromium. Chromium is by far the most critical of the elements needed to make stainless steel. Chromium promotes passivation and reduce corrosion rate in the passive state in various corrosive media. The amount of chromium required for passivity depends on the media and its concentration and temperature [1].

Nickel and manganese are both austenitizing elements, and to a certain degree are interchangeable in austenitic stainless steels with nickel being preferred

M.B. Abuzriba · S.M. Musa (✉)
University of Tripoli, Materials & Metallurgical Engineering Department, P.O. Box 13304 Tripoli, Libya
e-mail: sm_musa@yahoo.com

M.B. Abuzriba
e-mail: mokhtarphd@yahoo.com

E.K. Polychroniadis et al. (eds.), *2nd International Multidisciplinary Microscopy and Microanalysis Congress*, Springer Proceedings in Physics 164,
DOI 10.1007/978-3-319-16919-4_27

technically [2]. During the Korean War when there was a shortage of nickel, the AISI 200 series of austenitic stainless steels was developed. The 200-series stainless steels contain only half of the nickel found in the 300-series, plus the addition of manganese and nitrogen.

On the world market the price of manganese is cheaper Thannickel. Because of this, and the fact that the two metals can be used interchangeably to a certain degree, the use of manganese in the production of austenitic stainless steels should become more prevalent in the future.

For the reasons previously explained, the substitution of strategic and expensive materials has been and still is the subject of many articles. The authors of these articles urge researchers and alloy designers to explore the possibility of developing new austenitic stainless steel alloy systems that substitute for both chromium and nickel. Researchers developed a new family of alloys by replacing chromium with aluminum and nickel with manganese in the austenitic Fe–Cr–Ni system. These new alloys showed promise when initially evaluated.

The Fe–Al–Mn system was first established by koster and tonn [3]. These researchers studied the equilibrium phase diagram of Fe–Al–Mn alloy system using metallographic and dilatometric techniques. Their results indicate a stable austenitic (y) phase with up to 20 % Al and 50 % Mn, and they located the two-phase field of ferrite and austenite. Their results also showed that when aluminum and manganese are included in higher percentages, the austenitic phase was unstable, with a transformation to B-Manganese occurring. No indication of temperature or composition dependence of this transformation was made nor did they include it in their proposed phase diagram. Schmatz [4] later.

Generally confirmed and slightly modified the ternary system established by koster and tonn. He reported that the face centered cubic (FCC) structure of high aluminum and high manganese alloys was unstable with a transition to a brittle B-Manganese structure occurring on slow cooling. At a manganese content of 35 % only 7 % aluminum could be added if a complete FCC structure was to be obtained.

The first published work evaluating the mechanical properties of Al–Mn–Fe alloys was that by ham and cairn [5] the alloy used by ham and cairns contained 10.2 % Al, 34.4 % Mn, 0.76 % C, and the balance Fe. They called their substituted stainless steel "poor man's stainless" because of relative cheapness of aluminum. This alloy is about 15 % less dense than carbon steels or type 18–8 stainless steels, and about 20–25 % less dense that the nickel and cobalt-base super alloys. It has the ability to be strengthened via work hardening and strain aging. Ham and cairns presented a table comparing the results of mechanical tests on the substituted alloy to standard stainless steels. Banerji [6] surveyed all of one date reported on mechanical properties of the substituted alloys. In his paper, he drew a bar graph comparing the yield strength, tensile strength, reduction in area, and elongation of Ham and cairns alloy to standard steels.

Charles et al. [7] published work evaluating the mechanical properties of Al–Mn substituted stainless steels at low temperatures. The base alloy used contained 30 % Mn, 5 % Al, and the remainder Fe.

Garcia et al. [8] tested the oxidation of various Al–Mn–Fe alloys. Their alloys contained 7–13 % Al, 24–33 % Mn, 1–3 % Si, small amounts of C, Cu, P, S, and the remainder Fe. Homogenized samples were analyzed thermogravimetricaly. Experimental conditions; latm of flowing oxygen at 500 °C and at 700 °C.

The results at 700 °C for 24 h runs show that the behavior of the Al–Mn–Fe alloys except for one example behaved in a superior manner to carbon steel. One alloy composed of 12.4 % Al, 4.13 % Mn, 1.93 % Si, 0.03 % C, 8.9 % Cu, 0.02 % P, 0.015 % S, and the balance Fe, behaved similarly to carbon steel at 700 °C. Results for 500 °C evaluations indicate comparative or superior behavior to 304 stainless steel for all alloys. These researchers also found that the addition of Cu tended to increase the oxidation resistance of Fe–Al–Low Mn alloys at 500 °C.

The previous discussion of the mechanical properties and oxidation resistance of Fe–Al–Mn alloys indicate that these substituted alloys are superior or equivalent to the common stainless steels in these areas. However, there is no published work evaluating the corrosion properties of these substituted alloys.

2 Materials and Methods

Alloy Preparation These alloys were prepared for corrosion testing using an arc furnace with a protective atmosphere of argon, and using commercial iron, electrolytic manganese, pure aluminum, silicon and carbon. The alloys were melted into 75 g bar-shaped ingots. Table 1 gives the chemical composition of the heats A, B and C that were prepared in the laboratory.

Resulting microstructures for the heats were dendritic. To eliminate the segregation to dendritic microstructure, samples were given several homogenization heat treatments. The heats A, B, and C were then cold rolled to sheets approximately 0.75 mm thick followed by a heat treatment at 1020 °C (1868F), with subsequent quench.

An austenitizing heat treatment was conducted on all the heats at 1020 °C (1868 F) for 1 h followed by water quenching. The specimens were cut to a size suitable for the specimen holder of the corrosion tester. The specimens were then polished through 600-grit using sic paper. All of the specimens were prepared for the electrochemical corrosion tests by the procedures described in ASTM G5.

Corrosion Test Techniques The following electrochemical corrosion tests were conducted using EG and G Model 350A, corrosion measurement system, on all three heats, (A, B, & C).

Table 1 Chemical analysis of steels studied (weight percent, Bal Fe)

Alloy	Al	Mn	Si	C	P	S
A	9.05	33.30	1.41	0.88	0.05	0.02
B	8.63	28.6	2.68	0.69	0.05	0.02
C	7.25	24.04	2.06	1.01	0.05	0.02

1. Anodic polarization measurements were conducted as recommended by ASTM G5, in a solution prepared from reagent grade sulfuric acid and deionized water and adjusted to a PH of 2.4.
2. Cyclic polarization measurements for localized corrosion (pitting and crevice corrosion) were conducted substantially following ASTM G61 as specified.
3. Tafel extrapolation measurements.

The specimens were prepared as described in recommended practice G5. The test solution was prepared from reagent grade sulfuric acid and deionized water and adjusted to a pH of 2.4. One-half hour before specimen immersion, high purity nitrogen purging was started and continued with a flow rate of 200 cm/minute. A saturated calomel electrode (SCE) was used as a reference electrode with a salt bridge containing the test solution. One hour after the test specimen was immersed in the solution, a potentiodynamic sweep at a rate of 0.167 mV/s (0.6 V/h) was started at a potential that was approximately 250 mV negative to the corrosion potential and terminated when a potential 250 mV positive to the corrosion potential was reached.

Metallographic Examination Three different metallographic samples were taken from the three heats A, B and C (Table 1). The samples were ground, polished, and etched using a solution of 20 ml methanol, 20 ml HNO_3 and 4 ml HCI. The microstructure was observed using light microscopy at a magnification of 200x. They revealed austenitic microstructures.

3 Results and Discussions

Uhlig [9] has emphasized the dependence of corrosion behavior in materials on electrochemical potentials of the various phases. Corrosion can proceed if an electrochemical cell can be established between phases in the material. For this reason it is advantageous to have a single phase structure in corrosion inducing environments.

In the steels studied the manganese content varied from a high value of 33.3 % down to about 24 % to obtain a single phase austenitic structure. Specimens were heat treaded for 1-h at 1020 °C and then water quenched. Observation under an optical microscope revealed a single phase austenitic structure as shown in Figs. 1, 2 and 3. Published work evaluating mechanical properties and oxidation resistance of Fe–Al–Mn alloys indicates that these alloys are superior or equivalent to the 300 series austenitic stainless steels in most mechanical properties. However, there is no published work evaluating the aqueous corrosion resistance of these alloys.

The possibility that these austenitic alloys might possess useful corrosion resistance is of importance and interest because of the present dependence on chromium ports. The corrosion resistance of these alloys was evaluated in a week sulfuric acid solution adjusted to a Ph of 2.4 and also in 3.56 wt% NaCl solution. The results obtained from the above tests are tabulated and graphically presented.

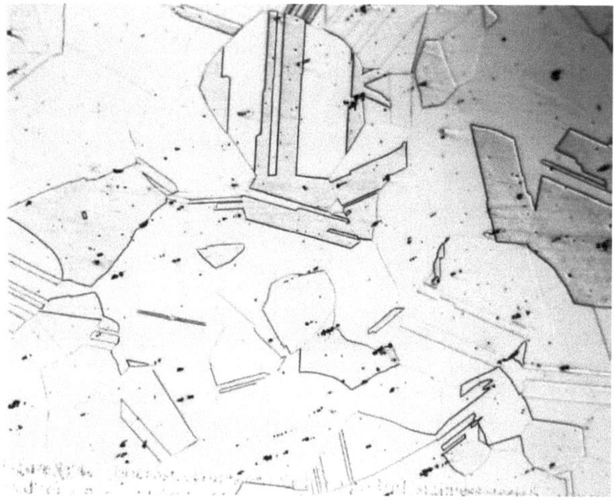

Fig. 1 Microstructures of specimen from heat A (solution heat treated for 1 h at 1020 °C, then water quenched). Magnification (200x)

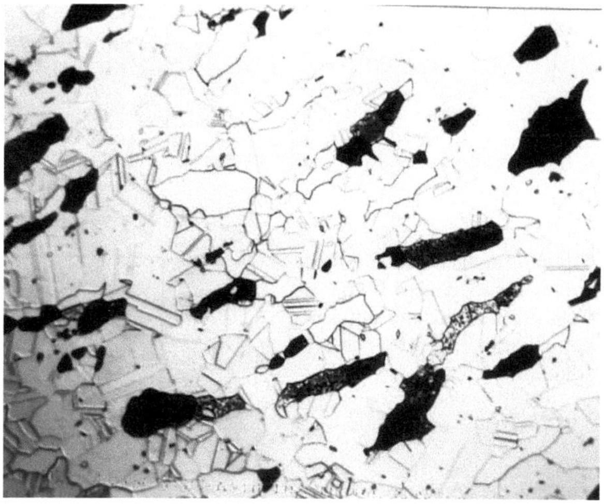

Fig. 2 Microstructures of specimen from heat B (solution heat treated for 1 h at 1020 °C, then water quenched). Magnification (200x)

The anodic polarization behavior diagrams were obtained from the corrosion measurement system. The results are summarized in Table 2 and presented in Fig. 4.

All the specimens from heats A, B, and C have very high critical current densities and high corrosion rates, as can be seen from Fig. 4. An active-passive transition was not observed in any of the specimens. The addition of aluminum up to 9 % and manganese up to 33 % were insufficient to cause an active to passive transition in the absence of chromium and nickel. The pitting behavior of the heats A, B, and C was evaluated by cyclic potentiodynamic polarization scans in 3.56 % NaCl solutions at ambient temperature. The heats A, B, and C exhibited great susceptibility to pitting corrosion as indicated by the results of the potentiodynamic

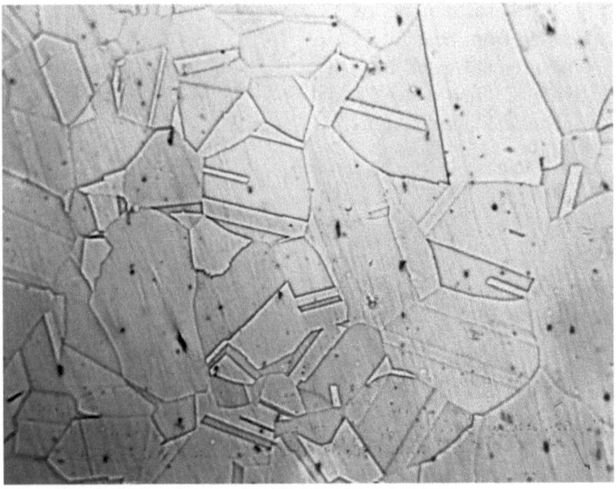

Fig. 3 Microstructures of specimen from heat C (solution heat treated for 1 h, then water quenched). Magnification (200x)

Table 2 Anodic polarization results

Heat	E corr (V vs. SCE)	I corr (NA/CM2)
A	−0.659	2.549×10^5
B	−0.650	1.818×10^5
C	−0.635	1.359×10^5

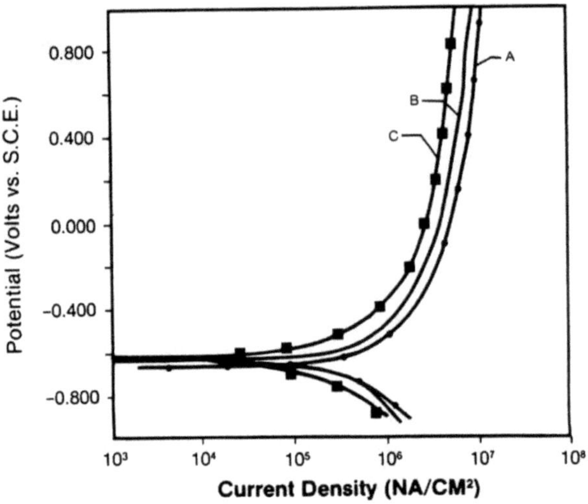

Fig. 4 Anodic polarization curves for specimens from heats A, B and C in a sulfuric corrosion in 3.56 wt% NaCl solution at 25 °C

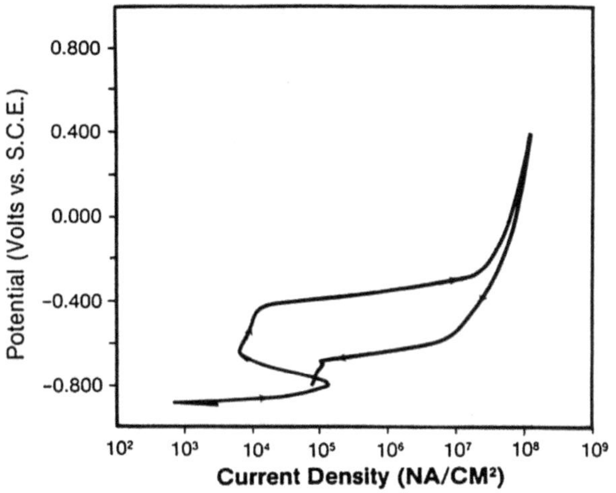

Fig. 5 Cyclic polarization curves for specimens from heat A that exhibited pitting acid solution adjusted to PH of 2.4 at 25 °C

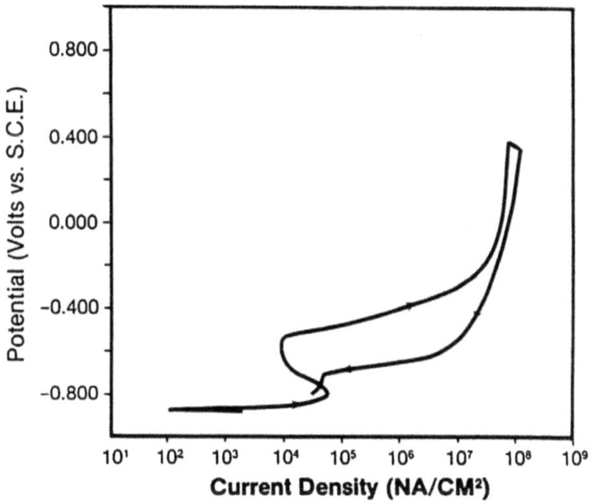

Fig. 6 Cyclic polarization of specimen from heat B that exhibited pitting corrosion in 3.56 wt% NaCl solution at 25 °C

polarization scans presented in Figs. 5, 6, 7 and 8 and the date obtained from these figures are summarized in Table 3.

The corrosion rates were reported in millimeters per year of penetration by using Tafel extrapolation technique. The corrosion rate obtained is reported in Table 4.

The values of the constants $ß_C$ and $ß_a$ were determined for each heat investigated by calculation from the Tafel plots. The unusually high values of Tafel slopes observed in some cases are attributed to apparent formation of a surface coating or to partial passivation. In some cases no linear Tafel region was observed and an estimated Tafel constant was obtained by the microprocessor of the testing instrument. It was possible to observe the surface of the specimen during the test. The point at which the coating formed was not determined precisely, but typically

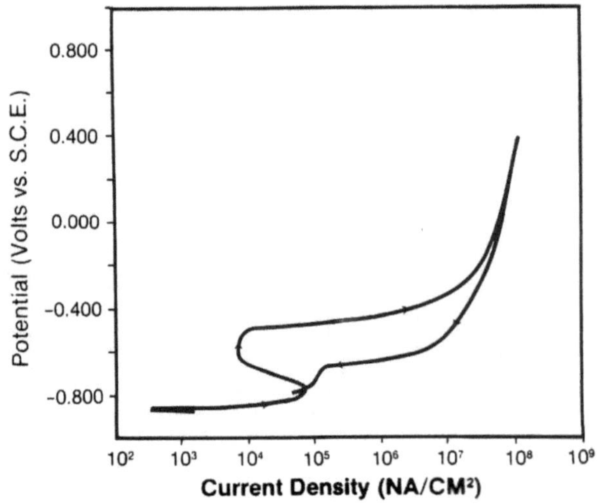

Fig. 7 Cyclic polarization of specimen from heat C that exhibited pitting corrosion in 3.56 wt% NaCl solution at 25 °C

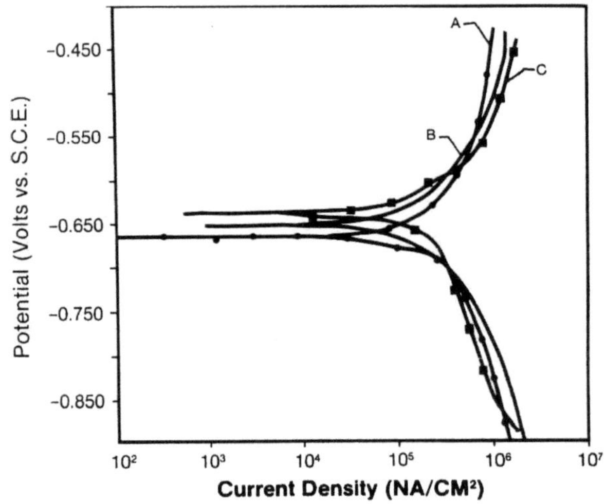

Fig. 8 Tafel plots for specimens from heats A, B and C in a sulfuric acid adjusted to pH of 2.4 at 25 °C

Table 3 Cyclic polarization results

Heat	E corr (v vs. SCE)	Eb (v vs. SCE)	Ep (v vs. SCE)	A (nA-V)
A	−0.879	−0.448	−0.750	10.8×10^6
B	−0.868	−0.530	−0.750	21.7×10^6
C	−0.860	−0.525	−0.770	8.6×10^6

Table 4 Tafel extrapolation results

Heat	ß$_C$	ß$_a$	Ecorr (v vs. SCE)	Icorr (NA/CM^2)	Corr. rate (MMPY)
A	0.300	0.180	−0.662	3.196×10^5	3.775
B	0.217	0.287	−0.651	2.179×10^5	2.573
C	0.270	0.195	−0.640	1.993×10^5	2.354

the coating would begin to fall away at about 200 mV anodic to Ecorr. It was originally believed that the aluminum content provided corrosion resistance to the alloys similar to that of chromium in stainless steels. Obviously aluminum content contributes nothing to the corrosion resistance of these alloys. As already discussed, Fe–Al–Mn alloys have mechanical properties and oxidation resistance comparable and sometimes superior to 18 % chromium-8 % nickel austenitic stainless steel. However, the corrosion resistance as tested and discussed seems to be rather poor and certainly not comparable to 18 % chromium-8 % nickel steels in the environments in which the tests were conducted.

4 Conclusions

The anodic polarization, cyclic polarization, Tafel extrapolation and linear polarization techniques were applied to study and evaluate the corrosion resistance of austenitic Fe–Al–Mn alloy steels in a sulfuric acid adjusted to a pH of 2.4 and in 3.56 wt% NaCl. Though the mechanical properties and the oxidation resistance of these alloys have been reported to be comparable or superior to 18 % chromium-8 % nickel austenitic stainless steels, the corrosion resistance was found to be poor and certainly not comparable to those of 18 % chromium-8 % nickel stainless steels.

References

1. R.A. Lula, Effects of material shortages on stainless steel technology. Met. Prog. **80**, 60–65 (1979)
2. R.M. Nadkarni, Case histories as a guide to future energy requirements for critical materials extraction systems proceedings, national symposium on critical and strategic materials, June. American chemical society, p. 108 (1978)
3. W. Koster, W. Tonn, The iron corner of the iron—manganese-aluminum system, archive fur das eisen hutten wesen, Vol. 7, p. 365
4. D.J. Schmatz, Formation of beta manganese type structure in iron—aluminum—manganese alloys. Trans. Am. Inst. Min. Metall. Eng. **215**, 112
5. J.L. Ham, R.E. Cairns Jr., manganese joins aluminum to give strong stainless steel. Prod. Eng. **29**(52), 50 (1958)
6. S.K. Banerji, An austenitic stainless steels without nickel and chromium. Met. Prog. **113**, 54–62 (1978)

7. J. Charles, A. Berghezan, A. Lutts, P.L. Dancoisne, New cryogenic materials: Fe-Mn-Al alloys. Met. Prog. **119**, 71–74 (1981)
8. J.C. Garcia, N. Rosas, Development of oxidation resistance Fe-Al-Mn alloys. Met. Prog. **122**, 47–50 (1982)
9. H. Uhlig, *Corrosion and Corrosion Control*, 2nd edn. (Wiley, New York, 1971)

Part II
Applications of Microscopy in the Biological Sciences

Ca, P and Collagen Fibrils Period Measurements in the Vertebras of Lordotic *Sparus aurata*

Panagiotis Berillis and Nikolaos Panagiotopoulos

Abstract Skeletal deformities of Gilthead seabream (*Sparus aurata*) is a major factor that effects the production cost, the external morphology of the fish as well as its survival and growth. Adult individuals of *S. aurata* were collected from a commercial fish farm and were divided into two groups. One with the presence of lordosis and one without any skeletal deformity. Fishes X-rayed and vertebras were taken from the site of the vertebra column that the lordosis occurred. One part was decalcified and prepared for collagen examination with transmission electron microscope and the rest incinerated and Ca and P contents were measured. The stoichiometries of the samples were obtained by EDS. The same procedure was followed for fish without any skeletal deformity (vertebras were taken from the middle part of the vertebra column). The decalcified vertebras parts examined with TEM, collagen micrographs were taken and the fibrils' period was measured. Statistics revealed no significant difference for both Ca and P or the collagen fibrils' period between the two fish groups.

1 Introduction

Gilthead seabream (*Sparus aurata*) is one of the most important farmed fish species in the Mediterranean region [1]. One of the major factor that affects the production cost is skeletal deformities. The development of these deformities is not well understood and related with nutritional, environmental and genetic factors [2].

P. Berillis (✉)
Department of Ichthyology and Aquatic Environment, School of Agricultural Sciences, University of Thessaly, Volos, Greece
e-mail: pveril@uth.gr

N. Panagiotopoulos
Department of Physics, University of Ioannina, Ioannina, Greece
e-mail: ntpanag@gmail.com

E.K. Polychroniadis et al. (eds.), *2nd International Multidisciplinary Microscopy and Microanalysis Congress*, Springer Proceedings in Physics 164,
DOI 10.1007/978-3-319-16919-4_28

A variety of abiotic factors such as environmental temperature, high swimming activity, photoperiod, tank colors, water temperature have proposed as significant in the development of skeletal deformities [3–7]. Nutritional factors such as vitamin A, enrichment of live food with DHA, retinoid treatment, D3 vitamin are also significant [8–12]. Skeletal deformities downgrade hatcheries' production, even in the case of well-studied species (*Dicentrarchus labrax* and *Sparus aurata*). Results from case studies of quality control in marine hatcheries during the last 15 years, show that skeletal deformities affect 7–20 % on average of the produced juveniles [13]. Sometimes these deformities can affect the 45–100 % of the total fish juveniles [13].

In bone, collagen represents more than 90 % of the organic bone matrix. It confers resistance to the structure and establishes the biomechanical properties of the tissue [14]. Fish bones consist of calcium-phosphor hydroxyapatite salts (inorganic part, about 65 % of bone's dry mass) embedded in a matrix of type I collagen fibers (organic part) [15]. The relationship between collagen and hydroxyapatite is crucial for bone toughness and stiffness. The collagen fibril diameter has been regarded as the most important factor related to biomechanical strength of tissues [16–18]. Regarding the bone mineral content, it is generally accepted that its measurement is an appropriate estimate of bone strength. The determination of the Ca and P levels may provide a sensitive measure of bone mineral changes and may add to our understanding of the changes occurring in bone diseases.

2 Materials and Methods

At the present study 30 adults individuals of *Sparus aurata* were collected from a commercial fish farm. Fishes were divided into two groups. One with the presence of lordosis (18 fishes, mean weight 267.57 ± 55.27 g, mean length 24.01 ± 1.30 cm), and one without any skeletal deformity (12 fishes, mean weight 247.69 ± 63.20 g, mean length 25.05 ± 1.82 cm). Each fish X-rayed at 50 KV (Fig. 1). The vertebra column of each *S. aurata* was divided in 3 parts. The cervical, with 4 vertebras, the abdominal, with 12 vertebras and the caudal with 8 vertebras.

Moreover, three vertebras were taken from the part of the vertebra column that the lordosis occurred. The two of them were incinerated at 600 °C and Ca and P contents were measured. The stoichiometries of the samples were obtained by Energy Dispersive Spectroscopy (EDS) in a Jeol JSM-6510LV SEM using a X-act Oxford Instruments X-ray analyzer. The third vertebra was carefully freed from soft tissue, and prepared for electron microscopy. It was fixed in a solution containing 4 mL of 0.05 M phosphate buffer (pH 7.2), 0.5 mL of 25 % glutaraldehyde, and 0.5 mL of 12.5 % ethylene-diamine tetraacetic acid (EDTA) disodium salt. After 2 h the bone was administered changes of 2 mL of EDTA in 3 mL of phosphate buffer until decalcified. Bone samples were trimmed into approximately 0.5 mm^3

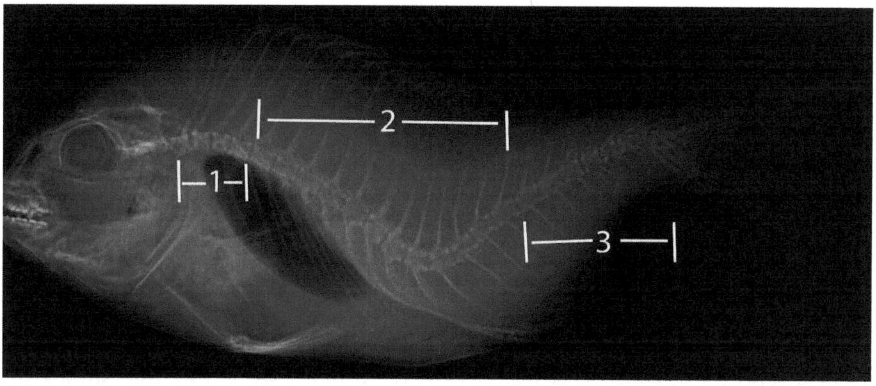

Fig. 1 X-ray of *Sparus aurata* a with lordosis deformity. The three parts of the vertebral column are shown: 1: cervical part, 2: abdominal part, 3: caudal part

cubes and fixed in 2.5 % glutaraldehyde in 0.05 M phosphate buffer (pH 7.2). Specimens were washed in the same buffer before and after fixation and rinsed in distilled water. Fixed specimens were then dehydrated in a graded alcohol series and impregnated overnight with 1:1 mixture of propylene oxide, and the resin used for embedding. Final embedding was performed in capsules with agar resin. Polymerization was completed in 48 h at 60 °C. Ultrathin sections (about 60 nm) were positively stained with 2 % aqueous solution of phosphotungstic acid (PTA; pH 3.3) for 1 h and then with 2 % aqueous solution of uranyl acetate (UA; pH 4.3) for 30 min, washing briefly in distilled water before and after UA staining. The same procedure was followed for fishes without any skeletal deformity.

Electron microscopy was performed on a Philips CM-10 TEM equipped with an Olympus Veleta digital camera. Micrographs were taken at ×19000–×92000. For the fibril periodicity measurements, areas of vertical-sectional collagen fibrils were photographed. A minimum of 104 collagen fibrils were analysed. In order to measure the fibril's D-period. Each fibril period was measured using the Olympus iTEM image-processing program. All values are presented as means ± standard deviation and differences presented at the 5 % level are considered significant. Collagen fibrils period values were checked for normality using the Shapiro–Wilk test. For the collagen fibrils period and the Ca, P levels the T-test was used.

3 Results and Discussion

The EDS analysis of the vertebras of both *S. aurata* groups showed that neither Ca nor P levels are effected by the lordosis deformity. Ca levels were respectively 65.87 ± 4.93 and 68.66 ± 6.03 wt% for individuals with lordosis and for the individuals without any skeletal deformity, while the P levels were 31.91 ± 3.88 and

Table 1 Ca, P, Ca/P levels and collagen fibrils mean period of *Sparus aurata* vertebras

	Ca level (wt%)	P level (wt%)	Ca/P ratio	Collagen fibrils mean period (nm)
Sparus aurata with lordosis	$65.87^{a} \pm 4.93$	$31.91^{b} \pm 3.88$	$2.13^{c} \pm 0.25$	$53.87^{d} \pm 7.25$ (n = 126)
Sparus aurata without any skeletal deformity	$68.66^{a} \pm 6.03$	$31.34^{b} \pm 6.00$	$2.36^{c} \pm 0.12$	$54.63^{d} \pm 3.95$ (n = 104)

Data are presented as means ± S.E. Means in a column followed by the same superscript are not significantly different ($p > 0.05$)

31.34 ± 6.00 wt%. Ca/P ratio was 2.13 ± 0.25 and 2.36 ± 0.12 for lordotic and normal fishes. The collagen fibrils mean period is similar for both groups (54.63 ± 3.95 and 53.87 ± 7.25 nm). Results are shown at Table 1.

Calcium and phosphorus are related with the development of the skeletal system and the stability of the vertebrae is maintained by a solid phase of calcium phosphate. The mineral content of the bones can be measured by many technics such as Auger electron spectroscopy, EDS or use of a monoenergetic, high photon flux X-ray beam [19]. Fishes absorb Ca and P from water. Ca requirement is met by their ability to absorb this element directly from water. Diet is the main source of P as the P concentration is low in seawater [20]. Calcium deficiency is not common in fish but phosphorus deficiency leads to reduced growth, reduced bone mineralization and skeletal abnormalities [21]. The ratio of Ca/P in the whole body of several fish species ranges from 0.7 to 1.6 [21]. Ozawa and Suzuki [22] refer that the Ca/P ratio in Japanese sea breams heat treated (800 °C) bones was 2.02 which is very close to our results.

The relationship between collagen and hydroxyapatite is crucial for bone toughness and stiffness. Mineralized collagen fibrils are the basic structural units of bone and many other calcified tissues. Histologically, fish bones are similar to those of higher vertebrates. Most of fish bones, both cellular and acellular, possesses relatively great amounts of uncalcified collagen fiber bundles [23]. In pathological bone tissue, deviations from normal fine structure are reflected in abnormal arrangements of collagen fibrils and abnormalities in fibril architecture [24]. In fish bones the major collagen type is type I [15]. The most striking feature of the fibrils is the regular transverse banding with its axial periodicity (D), where D is about 68 nm [25, 26]. Our collagen fibrils periodicity measurements are smaller than the 68 nm, but in electron microscopy dehydrated specimens usually gives lower values, than those obtained with low-angle X-ray diffraction of hydrated specimens and atomic force microscope where D is close to 68 nm [27] (Fig. 2).

In conclusion, Ca and P vertebras amounts and vertebras' collagen periodicity seem not to be the reasons of the lordosis formation in *S. aurata*. Further research, especially on collagen architecture, is needed.

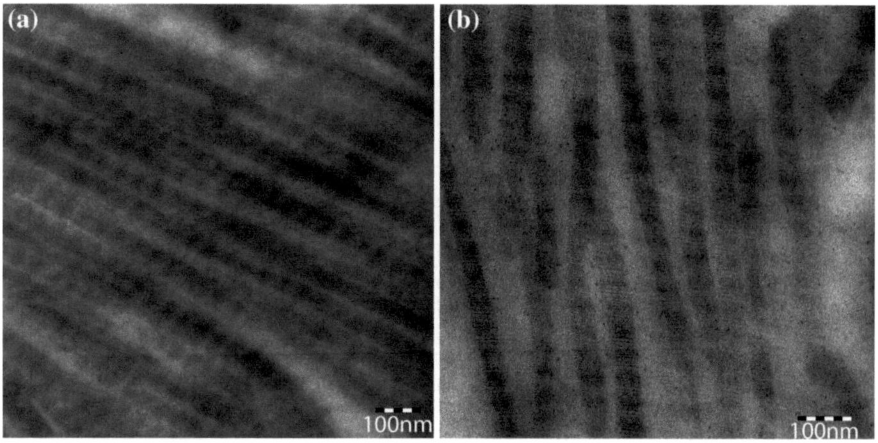

Fig. 2 **a** Collagen fibrils from the vertebra of *Sparus aurata* without any skeletal deformity. **b** Collagen fibrils from the vertebra of *Sparus aurata* with lordosis. At both images the collagen periodicity is obvious

Acknowledgments The authors would like to acknowledge the Electron Microscopy Facility of University of Ioannina for providing access to the EDS system.

References

1. FAO, Food and Agriculture Organization of the United Nations, Cultured Aquatic Species Information Program, Sparus aurata (Linnaeus, 1758) (2014), http://www.fao.org/figis/servlet/SQServlet?file=/work/FIGIS/prod/webapps/figis/temp/hqp_6595960067401687523.xml&outtype=html (accessed 14-5-2014)
2. I. Fernández, F. Hontoria, J.B. Ortiz-Delgado, Y. Kotzamanis, A. Estévez, J.L. Zambonino-Infante, E. Gisbert, Larval performance and skeletal deformities in farmed gilthead sea bream (*Sparus aurata*) fed with graded levels of Vitamin A enriched rotifers (*Brachionus plicatilis*). Aquaculture **283**, 102–115 (2008)
3. A. Tandler, F.A. Anav, I. Choshniak, The effect of salinity on growth rate, survival and swimbladder inflation in gilthead seabream, *Sparus aurata*, larvae. Aquaculture **135**, 343–353 (1995)
4. M. Kihara, S. Ogata, N. Kawano, I. Kubota, R. Yamaguchi, Lordosis induction in juvenile red sea bream, *Pagrus major*, by high swimming activity. Aquaculture **212**, 149–158 (2002)
5. A.J. Trotter, S.C. Battaglene, P.M. Pankhurst, Effects of photoperiod and light intensity on initial swim bladder inflation, growth and post-inflation viability in cultured striped trumpeter (*Latris lineata*) larvae. Aquaculture **224**, 141–158 (2003)
6. D.G. Sfakianakis, E. Georgakopoulou, I.E. Papadakis, P. Divanach, M. Kentouri, G. Koumoundouros, Environmental determinants of haemal lordosis in European sea bass, *Dicentrarchus labrax* (Linnaeus, 1758). Aquaculture **254**, 54–64 (2006)
7. J.M. Cobcroft, S.C. Battaglene, Jaw malformation in striped trumpeter *Latris lineata* larvae linked to walling behaviour and tank colour. Aquaculture **289**, 274–282 (2009)
8. T. Takeuchi, J. Dedi, Y. Haga, T. Seikai, T. Watanabe, Effect of vitamin A compounds on bone deformity in larval Japanese flounder (*P. olivaceus*). Aquaculture **169**, 155–165 (1998)

9. R.S.J. Gapasin, M.N. Duray, Effects of DHA-enriched live food on growth, survival and incidence of opercular deformities in milkfish (*Chanos chanos*). Aquaculture **193**, 49–63 (2001)
10. Y. Haga, T. Suzuki, H. Kagechika, T. Takeuchi, A retinoic acid receptor-selective agonist causes jaw deformity in the Japanese flounder, *Paralichthys olivaceus*. Aquaculture **221**, 381–392 (2003)
11. D. Mazurais, N. Glynatsi, M.J. Darias, S. Christodoulopoulou, C.L. Cahu, J.L. Zambonino-Infante, G. Koumoundouros, Optimal levels of dietary vitamin A for reduced deformity incidence during development of European sea bass larvae (*Dicentrarchus labrax*) depend on malformation type. Aquaculture **294**, 262–270 (2009)
12. M.J. Darias, D. Mazurais, G. Koumoundouros, N. Glynatsi, S. Christodoulopoulou, C. Huelvan, E. Desbruyeres, M.M. Le Gall, P. Quazuguel, C.L. Cahu, L. Zambonino-Infante, Dietary vitamin D3 affects digestive system ontogenesis and ossification in European sea bass (*Dicentrachus labrax*, Linnaeus, 1758). Aquaculture **298**, 300–307 (2010)
13. E. Georgakopoulou, P. Katharios, P. Divanach, G. Koumoundouros, Effect of temperature on the development of skeletal deformities in Gilthead seabream (*Sparus aurata* Linnaeus, 1758). Aquaculture **308**, 13–19 (2010)
14. L. Moro, M. Romanello, A. Favia, M.P. Lamanna, E. Lozupone, Posttranslational modifications of bone collagen type I are related to the function of rat femoral regions. Calcif. Tissue Int. **66**, 151–156 (2000)
15. J. Mahamid, A. Sharir, L. Addadi, S. Weiner, Amorphous calcium phosphate is a major component of the forming fin bones of zebrafish: indications for an amorphous precursor phase. Proc. Natl. Acad. Sci. **105**, 12748–12753 (2008)
16. D.A. Parry, G.R. Barnes, A.S. Craig, A comparison of the size distribution of collagen fibrils in connective tissues as a function of age and a possible relation between fibril size distribution and mechanical properties. Proc. R. Soc. London B Biol. Sci. **203**, 305–321 (1978)
17. G.H. Baek, G.J. Carlin, T.M. Vogrin, S.L. Woo, C.D. Harner, Quantitative analysis of collagen fibrils of human cruciate and meniscofemoral ligaments. Clin. Orthop. **357**, 205–211 (1998)
18. V. Ottani, M. Franchi, D. De Pasquale, L. Leonardi, M. Morocutti, A. Ruggeri, Collagen fibril arrangement and size distribution in monkey oral mucosa. J. Anat. **192**, 321–328 (1998)
19. M. Tzaphlidou, P. Berillis, D. Matthopoulos, Bone calcium, phosphorus detection by Auger electron spectroscopy. Micron **36**(7), 706–709 (2005)
20. S.P. Lall, P. Santosh, M. Lewis-McCrea Leah, Role of nutrients in skeletal metabolism and pathology in fish—an overview. Aquaculture **267**(1), 3–19 (2007)
21. S.P. Lall (2002) The minerals. In J.E. Halver, R.W. Hardy (eds.), Fish Nutrition, 3rd edn. (Academic Press Inc., San Diego, 2002), pp. 259–308
22. M. Ozawa, S. Suzuki, Microstructural development of natural hydroxyapatite originated from fish-bone waste through heat treatment. J. Am. Ceram. Soc. **85**(5), 1315–1317 (2002)
23. M.L. Moss, The biology of acellular teleost bone. Ann. N. Y. Acad. Sci. **109**(1), 337–350 (1963)
24. M. Tzaphlidou, Bone architecture: collagen structure and calcium/phosphorus maps. J. Biol. Phys. **34**(1–2), 39–49 (2008)
25. A. Miller, Molecular packing in collagen fibrils, in *Biochemistry of collagen*, ed. by G.N. Ramachandran, A.H. Reddi (Plenum Press, New York, 1976), pp. 85–136
26. B. Brodsky, E.F. Eikenberry (1982) Characterization of fibrous forms of collagen. Methods Enzymol. **82**(C), 127–174
27. M. Tzaphlidou, The role of collagen in bone structure: an image processing approach. Micron **36**(7), 593–601 (2005)

Use of Bone Marrow-Derived Mesenchymal Stem Cells in Improving Thioacetamide Induced Liver Fibrosis in Rats

Fatma A.A. Mansour, Iman Shaheed and Nabiha R.A. Hassan

Abstract Liver fibrosis, is one of big problems usually ends with cirrhosis which considered a life threatening disease as the only way of treatment is the liver transplantation, this study aimed to find a new way for fibrosis treatment by the use of bone marrow isolated Mesenchymal stem cells (MSCs). Thioacetamide (TAA) was used for fibrosis induction in male Sprague Dawely (SD) rats which divided into two random groups: group infused with TAA for fibrosis induction and group as control negative group. MSCs were isolated from bone marrow of twenty five (4–5) weeks male SD rats, and labeled with fluorescent material (PKH26) to confirm the homing of cells. After fibrosis induction, rats were divided into four subgroups to study the effect of MSCs injection in fibrosis treatment. After 4 weeks from MSCs administration, all rats were sacrificed. Liver tissue were collected for histopathological and immunohistopathological studies. In comparison with control groups, the treated groups with MSCs showed improvement in the amount of deposited collagen which decreased compared to control positive group. So MSCs can be used to replace liver transplantation in the treatment of fibrosis.

1 Introduction

Liver fibrosis is the healing response of the liver to chronic injury. Subsequent of repeated injury, the liver undergoes a tissue remodeling and ends with cirrhosis. It is characterized by excessive accumulation of extracellular matrix, with the formation of scar tissue encapsulating the area of injury [1]. The key finding was the identification of activated hepatic stellate cells (HSCs) as the major source of ECM in the fibrotic liver. Activation of HSCs is currently considered the critical common step in liver fibrosis. Although liver transplantation is a good alternative treatment, there are limited available donor livers for hundreds of millions of patients worldwide. So, it is

F.A.A. Mansour (✉) · I. Shaheed · N.R.A. Hassan
Department of Pathology, Faculty of Veterinary Medicine, Cairo University, Cairo 12613, Egypt
e-mail: Fatmaa_vet@yahoo.com

E.K. Polychroniadis et al. (eds.), *2nd International Multidisciplinary Microscopy and Microanalysis Congress*, Springer Proceedings in Physics 164,
DOI 10.1007/978-3-319-16919-4_29

very important to investigate appropriate therapies for the disease by different treatments. Mesenchymal stem cell (MSC) which is an adult type of stem cells is considered as an attractive cell source for regenerative medicine. MSCs have the capacity to differentiate into hepatocytes in vitro and in vivo. Further, MSCs administration could repair injured liver by reducing inflammation, collagen deposition and remodeling. So MSCs could not only have a possibility to repair acute damaged tissue but also have potentiality of reducing chronic fibrosis [2].

2 Materials and Methods

Preparing for BM-derived MSCs in rats as mentioned by [3, 4].

Fibrosis induction by using TAA Sixty male SD rats were acclimatized for 7 days and randomly divided into two groups, Group (A): include 50 rat, Injected intra-peritoneally with 200 mg\kg sterile thioacetamide (TAA) (Sigma, USA) with 5 % concentration twice weekly for 3 months [5]. Group (B): include 10 rats, as control negative group. The progression of fibrosis was demonstrated by scarification of three animals from group (A) and one from group (B) per month; in addition the mortality rate was 8 % in group (A). After 3 months, group (A) was subdivided into subgroups as described into Table 1. Then all animals were sacrificed after 1 month.

Clinical examination and Animal weighting the animals were regularly observed, and recording of any abnormal clinical signs. The animals were fastened overnight before the injection and the scarification time. Weighting with recording body weight before injection were done every week [6].

Postmortum and Histopathological Examination the sacrificed animals were subjected to careful postmortem examination. Liver tissues were fixed in 10 % buffered neutral formalin. Samples were routinely processed as paraffin embedded sections at 2–4 μ thickness; the prepared slides were stained by H&E, Masson trichrome stain and immunohistochemistery staining.

Fluorescent Microscope Examination Fluorescent microscope used for detection of marked stem cells in liver tissue to ensure homing of cells which were labeled with PKH26 [6].

Table 1 Shows the experimental design (effect of cells after TAA)

Main group	Sub-group	Animals no.	Treatment
(A)	(1)	10	Injected with TAA for 3 months only and treated with cells
	(2)	10	Injected with TAA for 4 months and treated with cells
	(3)	10	Control positive group
	(4)	7	Injected with TAA for 3 months only
(B)	(5)	7	Control negative group

Immunohistochemical Staining Immunohistochemical staining was performed to evaluate staining intensity of alpha-smooth muscle actin (α-SMA) [a protein marker for cirrhosis] in different groups by method described by sigma Aldrich manual [4, 6].

Statistical Analysis One way and two way (ANOVA) tests were used in the findings of the body weight evaluation. Significant Difference (SD) test was used to evaluate the significance between groups when ANOVA is significant. The significance level was set as P value ≤ 0.05 significant. Statistical analysis was Performed using SPSS version 16 [7].

3 Results

Isolation and culture of rat bone marrow derived MSCs at the third day of bone marrow incubation, elongated adherent cells were observed by microscopical examination. Extensive proliferation of cells was observed at the sixth day till the ninth day, at the following 3 days a relatively homogenous culture of the elongated cells was obtained, the population of cultured cells reached confluence within 2 weeks (70–80 % confluence). The cells appeared as fibroblasts in the morphology.

Clinical examination and animal weighting Animals of group (3) appeared lethargic, weak, dull, depressed with rough hair. The body weight changes are described in Tables 2 and 3.

Postmortem Examination Liver of group (3) appeared pale in color with numerous small nodules on the surface giving micronodular appearance and firm in consistency. Two animals from group (4) had liver with slightly nodular appearance and hard texture.

Histopathological Examination Group (A) at the end of the third month, revealed severe cytoplasmic vacuolation, individual cell necrosis of coagulative type and enlarged nuclei of hepatocytes. Proliferation of fibroblasts, few strands of collagen stained blue with MTS were bridging between central veins to surround

Table 2 Shows the mean body weight of group (A) and group (B) [mean ± SD] through the first 3 months

Month group	First month	Second month	Third month
Group (A)	148.50 ± 8.6^{bA}	113.99 ± 9.5^{cdB}	71.78 ± 5.4^{fB}
Group (B)	$155.16 \pm 2.^{defA}$	176.76 ± 3.7^{bcA}	185.74 ± 48.5^{abA}

Means with different letters (a, b, c, d, e, f) and (A, B) within the same column are significantly different at P value ≤ 0.05

Table 3 Shows mean ± SD of body weight in groups (1–5) at the end of the experimental study

Group	Group(1)	Group(2)	Group(3)	Group(4)	Group(5)
Mean ± SD	157.35 ± 5.3^{a}	113.96 ± 3.2^{a}	75.14 ± 6.2^{b}	113.92 ± 9^{a}	207.64 ± 2^{a}

Means with different letters (a, b, c, d, e, f) are significantly different at P value ≤ 0.05

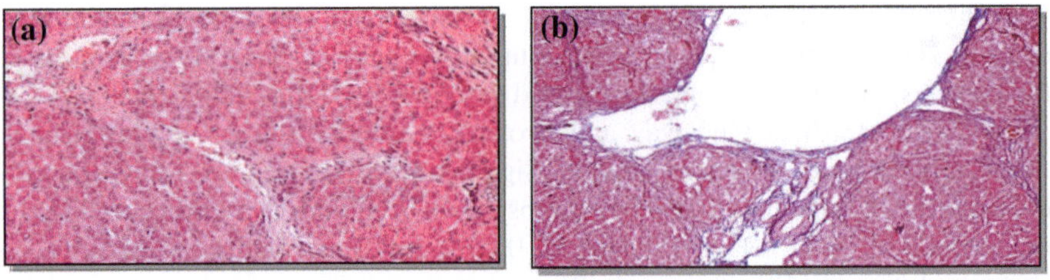

Fig. 1 **a** Liver of group (1) shows few strands of collagen with slight proliferation of fibroblasts between the hepatic lobules (H&E Stain X 200). **b** Shows few strands of collagen which stained *blue* in the portal area (MST Stain X 200)

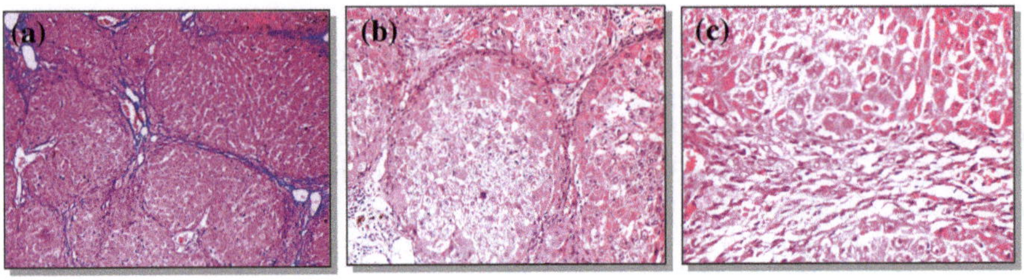

Fig. 2 **a** Liver of group (3) showing regenerative nodules appeared surrounded by *blue* stained collagen bundles (MST Stain X 200). **b** A regenerative nodule surrounded contain cells with vacuolar degeneration and most of cells were necrosed (H&E Stain X 400). **c** Showing apoptotic body (H&E Stain X 400)

the hepatic cells giving pseudolobulation appearance. The epithelial lining bile ducts showed hyperplasia with destructed corrugated basement membrane. Oval cells and Kupffer cells were highly activated. In between hepatocytes of Group (1) Liver, few strands of collagen were detected with slight proliferation of fibroblasts in five animals (Fig. 1a). In another four animals only few strands of collagen which stained blue with MTS were detected in the portal area (Fig. 1b). One animal showed complete pseudolobulation. In Group (2), only one animal showed few strands of collagen in the portal area. Six animals showed few strands of collagen between hepatocytes and bridging between central veins. The other three animals showed complete pseudolobulation as hepatic lobules were completely surrounded with thin collagen bundles. Liver of Group (3) showed multilobular cirrhosis, regenerative nodules appeared surrounded by collagen bundles stained blue with MTS (Fig. 2a). Each nodule contained cells with vacuolar degeneration and most of cells were necrosed (Fig. 2b). Others showed hyperatrophy with karyomegaly of the nucleus which appeared enlarged when compared with normal ones. Apoptotic bodies were observed as seen in (Fig. 2c). The main lesion in Group (4) was the vacuolation; few fibroblasts were proliferated in the portal area and in between hepatocytes. In four cases, few strands of collagen stained blue with MTS were observed bridging between central veins.

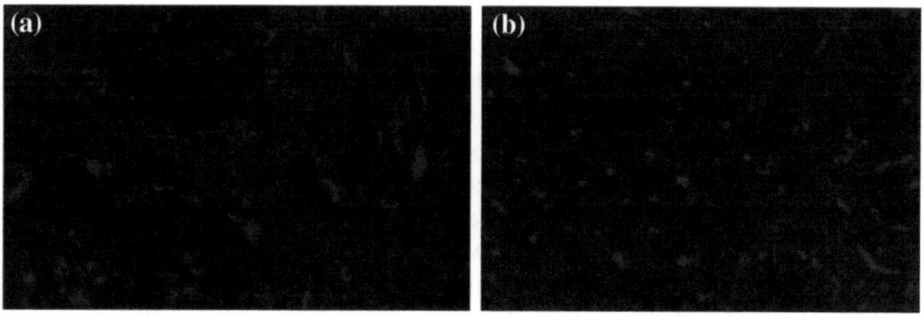

Fig. 3 **a** Unstained liver sections of rats treated with labeled PKH26 MSCs showing strong *red* autoflouresence between regenerative nodules and around blood sinusoids (200x) and **b** Showing diffuse pattern of strong *red* autoflouresence (200x)

Fluorescent microscope examination PKH26 stained cells showed strong red autoflouresence in unstained liver sections of animals treated with MSCs around blood sinusoids, in the fibrotic areas and between regenerative nodules in some animals (Fig. 3a). In other animals, PKH26 stained cells showed diffuse pattern as seen in (Fig. 3b).

Immunohistochemistery staining showed intense staining patterns in group (3), the intensity of α-SMA staining in TAA/MSCs groups [group (1) and group (2)] was lower when compared to control positive group.

4 Discussions

Animals treated with TAA appeared lethargic, weak, dull, depressed with rough hair [6]. Significant decrease in the body weight in animals treated with TAA could be attributed to the toxic effect of TAA decreasing the levels of nutrient absorption, energy utilization, and metabolic efficiency [8–10, 13]. While animals treated with MSCs showed significant increase in the body weight that could be contributed to the effect of MSCs. Micronodular appearance liver could be contributed to the development of collagen bundles which surround the regenerative nodules and cause pressure on the blood vessels, resulted in ischemia and decrease of the blood supply. So the color appeared pale grossly [4, 10].

The histopathological changes in the present study due to the administration of the TAA could be due to the metabolic conversion of TAA to free radical products: (thioacetamide sulfoxide) and (thioacetamide-S,S-dioxide) which attack microsomal lipids leading to their peroxidation, and production of reactive oxygen species (ROS), such as the hydrogen peroxide, super oxide anion and the hydroxyl radical. that result in liver injury and cirrhosis development [3, 8, 9, 11].

After treatment with MSCs, animals showed remolding of the collagen fibers which could be lysed by metalloproteinase (MMPs) more particularly the MMP2 that promote the degradation of (ECM) which have been secreted by MSC. The treatment effect of MSCs was higher in group (1) than (2). That as TAA injection was stopped in group (1) for 1 month during treatment with cells that was in agreement with [12].

PKH26 labeled MSCs showed strong red auto fluorescence after transplantation into rats, confirming that these cells were actually seeded into the liver tissue and differentiated into healthy cells replaced the damaged ones either hepatocytes or cholangiocytes which agreed with the work of [3, 4].

α-SMA expression showed intense staining patterns in cirrhotic animals, the intensity of α-SMA staining in TAA/MSCs groups were significantly less which indicate reductions in α-SMA expression reflecting a reduction in the number of activated HSCs by MSCs transplantation [4].

References

1. S.L. Friedman, Mechanisms of hepatic fibrogenesis. Gastroenterol. **134**, 1655–1669 (2008)
2. I. Sakaida, S. Terai, N. Yamamoto, K. Aoyama, T. Ishikawa, H. Nishina, K. Okita, Transplantation of bone marrow cells reduces CCl4-induced liver fibrosis in mice. Hepatol. **40** (6), 1304–1311 (2004)
3. Z.A. Amin, M. Bilgen, M.A. Alshawsh, H.M. Ali, A.H.A Hadi, M.A. Abdulla, Protective role of phyllanthusniruri extract against thioacetamide-induced liver cirrhosis in rat model, Evid.-Based Complement. Altern. Med. (2012)
4. J. Hussein, Z. El-khayat, E. Mostafa, L. Rashed, A. Farrag, D. Medhat, Mesenchymal stem cells therapy for thioacetamide induced liver cirrhosis. Int. J. Pharm. Pharm. Sci. **5**, 975–1491 (2013)
5. R. Bruck, O. Genina, H. Aeed, R. Alexiev, A. Nagler, Y. Avni, M. Pine, Halofuginone to prevent and treat thioacetamide-induced liver fibrosis in rats. Hepatol. **33**(2), 379–386 (2001)
6. M.T.A. Aziz, H.M. Atta, S. Mahfouz, H.H. Fouad, N.K. Roshdy, H.H. Ahmed, L.A. Rashed, D. Sabry, A.A. Hassouna, N.M. Hasan, Therapeutic potential of bone marrow-derived mesenchymal stem cells on experimental liver fibrosis. Clin. Biochem. **40**, 893–899 (2007)
7. G.W. Snedecor, W.G. Cochran., *Factorial experiments* (Iowa State University, 1980 in press) pp. 298–333
8. M.A. -Alshawsh, M.A. Abdulla, S. Ismail, Z.A. Amin, Hepatoprotective effects of Orthosiphonstamineus extract on thioacetamide-induced liver cirrhosis in rats. Evid.-Based Complement. Altern. Med. (2011)
9. C. Anbarasu, B. Rajkapoor, K.S. Bhat, J. Giridharan, A.A. Amuthan, K. Satish, Protective effect of Pisonia aculeate on thioacetamide induced hepatotoxicity in rats. Asian. Pac. J. Trop. Biomed. **2**(7), 511–515 (2012)
10. C. Ries, V. Egea, M. Karow, H. Kolb, M. Jochum, P. Neth, MMP-2, MTI-MMP and TIMP-2 are essential for the invasive capacity of human mesenchymal stem cells: differential regulation by inflammatory cytokines. Blood **109**, 4055–4063 (2007)
11. K. Dai, J.Y. Qi, D.Y. Tian, Leptin administration exacerbates thioacetamide-induced liver fibrosis in mice. World J. Gastroenterol. **11**(31), 4822 (2005)
12. A. Pellicoro, P. Ramachandran, J.P. Iredale, Reversibility of liver fibrosis. Fibrogenesis & tissue repair, **1**, pp. 5–26 (2012)
13. G. Poli, J.R. Schaur, 4-Hydroxynonenal in the pathomechanisms of oxidative stress. IUBMB Life **50**(4–5), 315–321 (2000)

Electrochemical Detection of Nicotine Using Cerium Nanoparticles Modified Carbon Paste Sensor and Anionic Surfactant

A.M. Fekry, S.M. Azab, M. Shehata and M.A. Ameer

Abstract A promising electrochemical sensor for the determination of nicotine (NIC) was developed by electrodeposition of Ce-Nanoparticles on a carbon paste electrode (CPE). The interaction of nicotine was studied using cyclic voltammetry (CV), electrochemical impedance spectroscopy (EIS), Scanning electron microscope (SEM) and Energy Dispersive X-Ray Analysis (EDX) techniques, in both aqueous and micellar media. The NIC Measurements were carried out in Britton–Robinson (B–R) buffer solution of pH range (2.0–8.0) containing (1.0 mM) sodium dodecylsulfate (SDS). The linear response range of the sensor was between 8×10^{-6} and 10^{-4} M with a detection limit of 9.43×10^{-8} M. Satisfactory results were achieved for the detection of NIC in real samples as urine and different brands of commercial cigarettes.

1 Introduction

Nicotine, 3-(1-methyl-2-pyrrolidinyl) pyridine, is the main alkaloid in tobacco leaves (Nicotiana tabacum L., Solanaceae). It is being best known for their use in cigarettes rather than for their therapeutical applications [1–3]. It has some high potential health hazard such as being the cardiovascular and respiratory disorder

A.M. Fekry · M. Shehata · M.A. Ameer (✉)
Chemistry Department, Faculty of Science, Cairo University, Giza 12613, Egypt
e-mail: mameer_eg@yahoo.com

A.M. Fekry
e-mail: hham4@hotmail.com

M. Shehata
e-mail: mohammed.shehata9011@yahoo.com

S.M. Azab
Pharmaceutical Chemistry Department, National Organization for Drug Control and Research (NODCAR), Giza-29, Egypt
e-mail: sheryspecial@yahoo.com

E.K. Polychroniadis et al. (eds.), *2nd International Multidisciplinary Microscopy and Microanalysis Congress*, Springer Proceedings in Physics 164,
DOI 10.1007/978-3-319-16919-4_30

including lungs cancer. The nicotine content of tobacco varies with tobacco type, but the normal nicotine level range goes from 1 to 3 % [1, 4]. Therefore, its determination is important in medicine, toxicology and tobacco industry. Over 20 years, spectrophotometry, chromatography, reversed HPLC, and potentiometry are development methods for nicotine determination. Some of these methods require preliminary extraction and purification of nicotine from the sample matrix and skilled personnel manipulating sophisticated instrumentation, [3] leading to considerable losses of the analyte to be determined. From the electrochemical point of view, very few studies have been made on this compound. This is mainly because the nicotine oxidation/reduction processes occur at extremely positive/negative potentials, which are out of the potential window of the conventional electrodes [1]. In this work, we introduce a sensitive electrochemical procedure for NIC detection based on Cerium nanoparticles modified carbon paste electrode (CNMCPE) that can overcome this obstacle. Carbon paste electrode has been widely applied in the electroanalytical community due to its good characteristics [5]. Electrodeposition of Ce-nanoparticles on CPE surface was an important method for detection of NIC. Ce nanoparticles showed a large participation in the area of electrochemical sensors, biosensors [6] and batteries [7]. Surfactants have been widely used in chemistry particularly affecting several electrochemical processes, as well as in the electroanalysis of biological compounds and drugs [1]. Carbon paste electrodes interact with surfactants, namely through surface adsorption. Thus, carbon paste electrode modified with surfactants proved to be useful for the determination of both inorganic species and biological compounds [8]. The use of surfactant solutions as modifiers can improve the sensitivity and selectivity of the voltammetric measurements [1].

The aim of this work is to construct a novel and sensitive electrochemical sensor based on Ce nanoparticles, CP and SDS to be used for the determination of nicotine in urine samples and cigarettes.

2 Experimental Work

Nicotine standard samples (99 %) were provided by Egyptian Eastern Company for Smoking and were used as received without prior purification. NIC stock solutions of about 1.62 g/L were freshly prepared in water then stored in a dark container. B–R (4.0×10^{-2} M) buffer solution of pH 2.0–8.0 ($CH_3COOH + H_3BO_3 + H_3PO_4$) were used for preparing the standard solutions of nicotine. pH values were adjusted using 0.2 M NaOH. Sodium dodecylsulfate (SDS) was purchased from sigma Aldrich. All solutions were prepared from analytical grade chemicals and deionized water. Cigarette samples were purchased from the local supermarkets. Two products of different international cigarette brands: L&M® and Marlboro® were chosen. All experiments were carried out at the room temperature. CPE with a diameter of 3 mm was prepared using few drops of paraffin oil mixed with graphite powder (0.5 g) and the mixture was blended thoroughly to obtain a homogeneous paste [9].

The resulting mixture was used to fill into a Teflon tube and pressed tightly. The CPE was polished with ultra-fine emery paper until a smooth surface was achieved. CPE surface was pre-treated by applying a potential of +1.3 V for 30 s in the blank supporting electrolyte without stirring, in order to increase the hydrophilic properties of the electrode surface through introduction of oxygenated functionalities, accomplished with an oxidative cleaning. Ce nanoparticles modified CPE was formed by dipping the clean CPE into 0.1 mM ammonium ceric nitrate solution, about 2 mL of lactic acid and the solution pH was adjusted using (33 %) ammonium hydroxide solution and then scanning in the potential range from −2.0 to +2.0 V at 25 mV/s for 10 cycles at (25 mL) glass cell [10].

All electrochemical and EIS measurements were performed with a personal computer-controlled electrochemical workstation IM6e Zahner-elektrik, GmbH, (Kronach, Germany). All EIS experiments were recorded between 0.1 Hz and 100 kHz with an excitation signal (sinusoidal potential) of 10 mV amplitude and the data analysis software was provided with the instrument and applied a nonlinear least-squares fitting. A typical three-electrode fitted with a large platinum sheet of size 15 × 20 × 2 mm as a counter electrode (CE), saturated calomel electrode (SCE) as a reference electrode (RE) and CNMCPE as the working electrode (WE) were used. A HANNA 213 pH meter with glass combination electrode was used for pH measurements. SEM measurements were carried out using SEM Model Quanta 250 FEG attached with EDX Unit (Energy Dispersive X-ray Analyses), with accelerating voltage 30 kV, magnification14x up to 1,000,000 and resolution for Gun.1n) (FEI company, Netherlands). All the electrochemical experiments were performed at an ambient temperature of 25 ± 2 °C.

Cigarettes were taken out of their rolling paper and dried at 40 °C oven for 30 min before weighing. A mixture of ten cigarettes taken from two packs of the same brand was placed (0.1 g) in a 50 mL glass vial with a cap. After adding 10 mL water using a pipette, the contents of the vial were sonicated for 3 h in an ultrasonic water bath and then filtered. 100 μL of the clear filtrate were mixed with the B–R buffer (pH 2.0) containing 1.0 mM sodium dodecylsulfate in the voltammetric cell and analyzed under same conditions as used to obtain calibration graph. Standard NIC was dissolved in urine (diluted 400 times using 100 mL of B–R buffer pH 2) to make a stock solution. Standard additions of urine were carried out from a solution containing NIC in 23 mL of B–R buffer pH 2 and 1.0 mM sodium dodecyl sulfate in the voltammetric cell and analyzed under same conditions as used to obtain calibration graph.

3 Results and Discussion

The SEM of the CPE and CNMCPE are shown in Fig. 1a, b, respectively. Significant differences in the surface structure were observed. The surface of the CPE was characterized by isolated and irregularly shaped graphite flakes and separated layers (Fig. 1a). The SEM image of CNMCPE (Fig. 1b) shows that

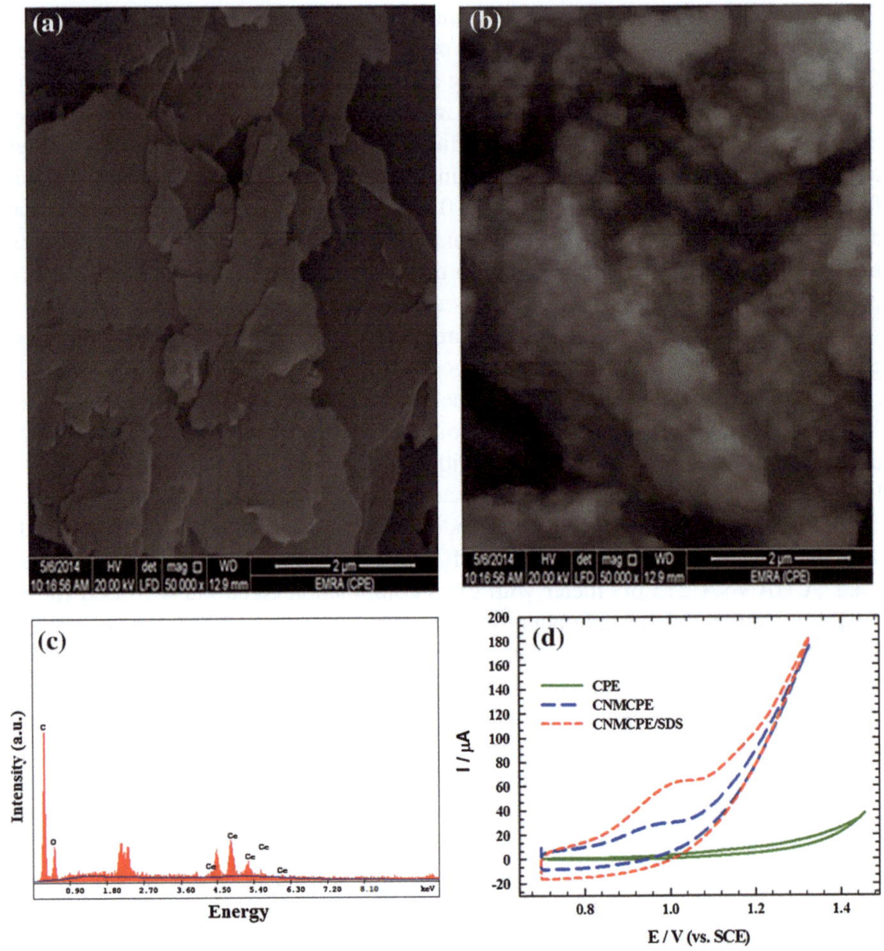

Fig. 1 **a** SEM image of CPE, **b** SEM image of CNMCPE, **c** EDX spectra of CeO_2 nanoparticles, and **d** CVs of 1.0×10^{-3} M NIC in B–R buffer pH 2.0 at a scan rate of 25 mV s^{-1} recorded at: bare CPE, CNMCPE and CNMCPE/1.0 mM SDS

metallic nanoparticles are located at different elevations over the substrate. Moreover, a random distribution and interstices among the nanoparticles were observed in the SEM image of the CNMCPE exhibiting a large random surface area. The EDX data of the CeO_2 nanoparticles is shown in Fig. 1c, indicates the presence of Ce and O peaks, and confirms that CeO_2 nanoparticles were indeed coated on the electrode surface. These results suggest that CeO_2 nanoparticles were successfully prepared by this simple electrochemical method. On the other hand, Fig. 1d shows typical CVs of 1.0×10^{-3} M NIC in B–R buffer pH 2.0 at a scan rate of 25 mV s^{-1} recorded at three different electrodes (i.e., (1) bare CPE, (2) CNMCPE and (3) CNMCPE/1.0 mM SDS. At bare CPE there is no oxidation peak observed, while the electrodeposition of Ce-nanoparticles onto the surface of the CPE (CNMCPE) was an effective strategy to enhance the detection of NIC, where CPE represents the best electrode type for Ce-nanoparticles electrodeposition [11]. Cerium dioxide (CeO_2) is the most stable oxide of cerium. This compound is also called ceria or ceric oxide [10]. Micellar media may induce two important effects on the electrochemical reactions. The first effect is that the surfactant can stabilize radical ions and other reaction intermediates, which effect on the mechanism of the electrode reaction. Secondly, the presence of surfactant molecules modifies the double layer structure [1]. The sensitivity of SDS modified electrode (CNMCPE/SDS) for the determination of NIC was clearly increased being an indicative that the rate of electron transfer was increased. The sensitivity of surfactant-modified electrode was about 2.45 times higher than CNMCPE. At pH 2.0, nicotine has a positively charged monoprotonated form [12]. On the other hand, the adsorption of the SDS micelles onto electrode surface forms a negatively charged hydrophilic film oriented towards the water bulk phase. According to this fact, positive charged nicotine has a tendency to accumulate in the negatively charged crown of anionic SDS micelles, which may increase the electron transfer rate [13].

The effect of solution pH on the electrocatalytic oxidation of NIC at the CNMCPE/SDS was studied by the CV technique using B–R buffers within the pH range of 2.0–8.0 (Fig. 2a). The increase in the solution pH shifts the anodic peak potentials to negative values, indicating that the electrocatalytic oxidation of NIC at the CNMCPE/SDS is a pH-dependent reaction. This shows that protons have taken part in their electrode reaction processes [14]. The highest oxidation peak current was obtained at pH 2 (inset), then the peak current decreased from pH 2.0 to pH 5.0 and increased again up to pH 8.0. Due to the weak diacidic base character of nicotine, it has two pK_a values, $pK_{a1} = 8.02$ and $pK_{a2} = 3.12$ which correspond to the protonation of pyrrolidine nitrogen (monoprotanated form) and pyridine nitrogen (diprotonated form) present in nicotine molecule, respectively [12]. These prove that the oxidation step of nicotine is located on the pyrrolidine ring and attributed to the oxidation of tertiary nitrogen.

Cyclic voltammetric behavior was examined further as a function of scan rate in order to obtain additional information regarding the adsorptive properties of the reaction studied. A plot of I_{pa} versus $\upsilon^{1/2}$ (υ ranging from 10 to 100 mV s^{-1}) for 1.0×10^{-3} M NIC using CNMCPE/SDS in B–R buffer (pH 2.0) gave a straight line relationship. The linearity of the relationship indicated that the electrochemical process was under diffusion control and that the adsorption of aggregates at the electrode surface was also possible. Typical CVs of NIC at different scan rates were shown in Fig. 2b. The oxidation peak currents increased linearly (inset) with the regression coefficient ($r^2 = 0.9997$ and $\gamma = 5$) suggesting that the reaction is diffusion-controlled electrode reaction. The relation between anodic peak current and scan rate has been used for the estimation of the "apparent" diffusion coefficient, D_{app}, for the compounds studied. D_{app} values were calculated from the Randles Sevcik equation [15]. D_{app}, of NIC using CNMCPE and CNMCPE/SDS in B–R buffer (pH 2.0) were calculated from cyclic voltammetry (CV) experiments and were found to be 1.177×10^{-6} and 7.35×10^{-6} cm^2 s^{-1}, respectively; this indicate

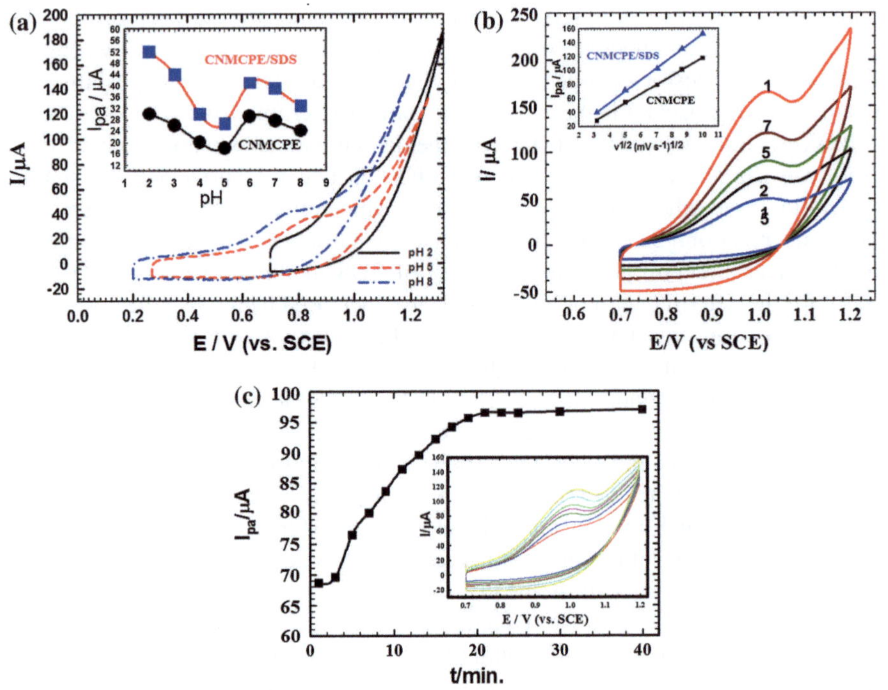

Fig. 2 CVs response of 1.0×10^{-3} M NIC at the CNMCPE/SDS **a** using B–R buffers of different pH values (*Inset* variation of the anodic peak current (I_{pa}) against pH using CNMCPE and CNMCPE/SDS), **b** at different scan rates using B–R buffer with pH 2 (*Inset* variation of I_{pa} with the square root of the scan rate ($\upsilon^{1/2}$) using CNMCPE and CNMCPE/SDS), and **c** variation of I_{pa} for 1.0×10^{-3} M NIC in B–R buffer pH 2.0 at the CNMCPE/SDS with accumulation times (typically between 1–40 min) (*Inset* CVs for 1.0×10^{-3} M NIC in B–R buffer (pH 2.0) solution recorded every 3 min)

that the redox reaction of the analyte species takes place at the surface of the electrode under diffusion control of the molecules from solution to the electrode surface.

In order to investigate the response of CNMCPE/SDS, the CV for 1.0×10^{-3} m NIC in B–R buffer (pH 2.0) solution were recorded every 3 min (Fig. 2c). It was found that the anodic peak currents increased with increasing the accumulation time. The highest current response for CNMCPE was reached after 21 min at 1.0 V versus SCE, so the optimum time for electrode stability is 21 min. Repetitive measurements indicated that this electrode has good reproducibility and stability during the voltammetric measurements.

Electrochemical impedance spectroscopy (EIS) is an effective tool for studying the interfacial properties of surface-modified electrodes [16]. The impedance plots are shown as both Bode (Fig. 3a, b) and Nyquist plots (Fig. 4a, b), respectively. The Nyquist plots include a small semicircle corresponding to charge transfer resistance and a linear part corresponding to diffusion process. The impedance data were thus

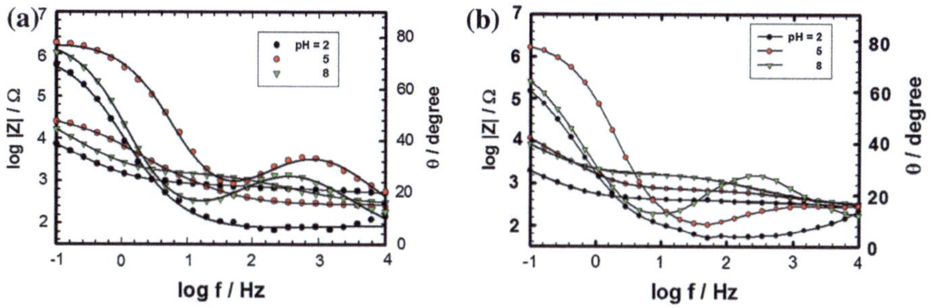

Fig. 3 The typical impedance spectrum represented in the form of a Bode plot of NIC using the **a** CNMCPE, **b** CNMCPE/SDS at different pH values (*Symbols* and *solid lines* represent the experimental measurements and the computer fitting of impedance spectra, respectively)

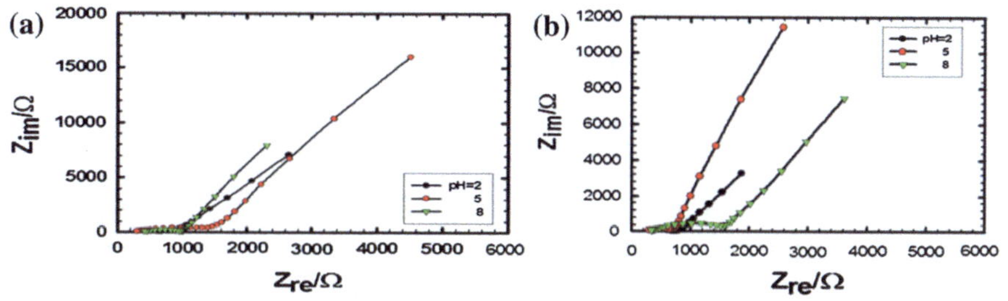

Fig. 4 The typical impedance spectrum represented in the form of a Nyquist plot of NIC using **a** CNMCPE, **b** CNMCPE/SDS at different pH values (*Symbols* and *solid lines* represent the experimental measurements and the computer fitting of impedance spectra, respectively)

simulated to the appropriate equivalent circuit for the cases exhibiting two time constants (Scheme 1). The appropriate equivalent model used to fit the high and low frequency data consists of two circuits in series from Z_wCPE_1 and $R_{CT}CPE_2$ parallel combination and both are in series with R_s. CPE_1 is related to the constant phase element (CPE) of the inner layer and CPE_2 to the outer layer, while R_{CT} is the charge transfer resistance of the outer layer related to the small semicircle at high frequency [17]. Z_w is a Warburg component related to the linear region at the lower frequencies in the Nyquist plot which is related to diffusion phenomena [18]. An empirical exponent ($\alpha = 0$–1) is introduced to account for the deviation from the ideal capacitive behavior due to surface roughness [19]. An ideal capacitor corresponds to $\alpha = 1$ while $\alpha = 0.5$ becomes the CPE in a Warburg component [18]. In all cases, good conformity between theoretical and experimental results was obtained with an average error of 1 %.

The best fitting values calculated from the equivalent circuit for the impedance data at the two different cases for the CNMCPE were collected. The value of solution resistance, R_s, for each pH was almost constant within the limits of the experimental errors. CNMCPE/SDS shows relatively higher values for the

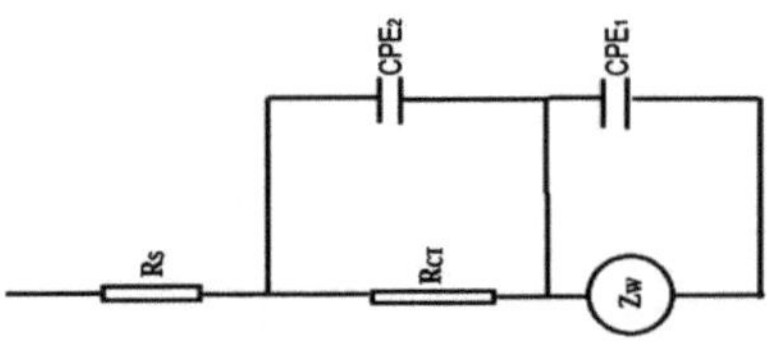

Scheme 1 The equivalent circuit used in the fit procedure of the impedance spectra

capacitance or lower values for the impedance compared to CNMCPE indicating a more conducting behavior due to electrode surface ionic adsorption by SDS. Also the same is observed for Z_W values, where it has higher values for CNMCPE/SDS than for CNMCPE. The lowest R_{CT} or highest CPE values obtained at pH = 2 for both tested electrodes indicated higher conductivity and confirming highest oxidation peak current obtained from CV's results. Z_W is characterized by a very low frequency slope of −0.5 and it intercept on the log Z axis at $f = 1$ Hz of $\sigma\pi^{-1/2}$, where σ is the Warburg impedance coefficient (ohm cm^2 s$^{-1/2}$) [20]

$$\log Z = \log \sigma\, \pi^{-1/2} - \frac{1}{2}\log f \tag{1}$$

σ can be obtained from (1) by getting log Z values at f = 1, σ values are evaluated as 2733.95 for CNMCPE and 1060.66 for CNMPE/SDS. Diffusion coefficient at pH = 2 can be calculated using the following equation [21]:

$$\mathrm{D} = \left[RT/\sqrt{2}AF^2\sigma C\right]^2 \tag{2}$$

where D is the diffusion coefficient, A is the area of the electrode, σ is Warburg coefficient, C is the concentration of nicotine, R is the gas constant, T is the temperature and F is Faraday constant. D is found to be equal 0.95×10^{-6}, for CNMCPE and 6.312×10^{-6} for CNMCPE/SDS at pH = 2. Thus diffusion coefficient values calculated from EIS measurements is comparable to that obtained from CV's measurements. This confirms well CV's data indicating that the redox reaction of the analyte species takes place under diffusion control process.

To prove the sensitivity of the CNMCPE toward the electrochemical measurement of NIC, the effect of changing the concentration of NIC in B–R buffer pH 2.0 and scan rate of 10 mV s^{-1} using square wave mode was studied (Fig. 5a). The corresponding calibration plot is given in Fig. 5a (inset). The peak current was evaluated as the difference between each voltammogram and the background electrolyte voltammogram. The calibration plot was linearly related to the NIC concentration over the ranges of 4.0×10^{-6}–5.0×10^{-4} M with correlation coefficient of 0.999. The concentration of NIC was calculated from the linear regression equation of the standard calibration curve: I_{pa} (μA) = 0.023 C (M) + 0.5886. The limits of detection (LOD) and the limits of quantitation (LOQ) were calculated using the following equations:

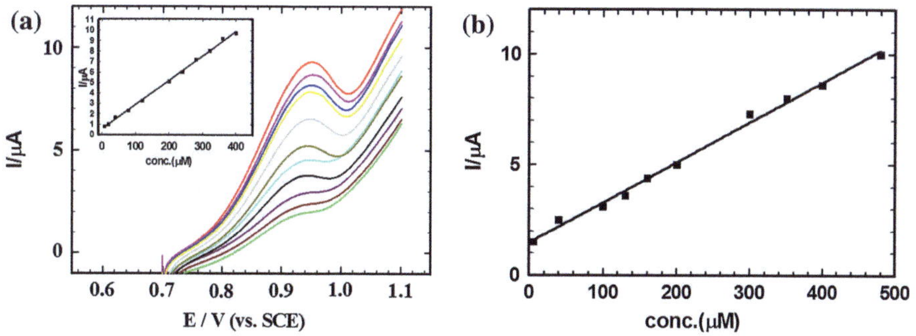

Fig. 5 **a** The CVs obtained at CNMCPE/SDS for successive addition of NIC and 1.0 mM SDS in B–R buffer pH 2.0 and a scan rate of 10 mV/s. (*Inset* the corresponding calibration curve of NIC) and **b** represent the calibration curve of NIC in urine

Table 1 Evaluation of the accuracy and precision of the proposed method for the determination of NIC in urine sample

[NIC] added (M) $\times 10^{-6}$	[NIC] Found[a] (M) $\times 10^{-6}$	Recovery (%)	SD $\times 10^{-6}$	S.E[b] $\times 10^{-6}$	C.L.[c] $\times 10^{-6}$
12.00	12.03	100.2	0.05	0.02	0.08
60.00	60.20	100.3	3.61	0.18	0.58
120.0	119.77	99.80	0.22	0.11	0.35
300.0	299.25	99.75	1.70	0.85	2.77

[a]Mean for five determinations
[b]Standard error = $SD/n^{1/2}$
[c]Confidence at 95 % confidence level

$$\mathrm{LOD} = 3s/b \tag{3}$$

$$\mathrm{LOQ} = 10s/b \tag{4}$$

where s is the standard deviation of the oxidation peak current (three runs) and b is the slope (μA/m) of the related calibration curves, and they were found to be 9.43×10^{-8} and 3.14×10^{-7} M respectively.

Validation of the procedure for the quantitative assay of NIC in urine was examined in B–R buffer, pH 2.0, at a scan rate of 10 mV s^{-1} (Table 1). The calibration curve (Fig. 5b) gave a straight line in the linear dynamic range of 6×10^{-6}–4.8×10^{-4} M. The concentration of NIC in urine samples was calculated from the linear regression equation of the calibration curve: I_{pa} (μA) = 0.0181C (M) + 1.4796, the correlation coefficient r^2 0.997, the LOD is 2.06×10^{-7} M and the LOQ is 6.89×10^{-7} M.

Two products of different cigarette brands were analyzed (L&M and Marlboro), to test the practical use of the procedure developed. The collapse of the bubbles created by the sonication of solutions results in the generation of high local energy and a high contact between solvent and solute, and can thus greatly increase the sample extraction efficiency. The sample preparation procedure used in this study was a slight modification of the method described by Suffredini et al. [14]. In our study, the ultrasonic extraction time of 3 h was chosen instead of 1 h in order to achieve a quantitative recovery of nicotine into aqueous solution. The results obtained in the analysis of different cigarette samples are shown in Table 2. Nicotine percentage values obtained are precise for all types of cigarette analyzed. It is necessary to indicate that, in practice, the nicotine reported in the cigarette packs correspond to the amount of nicotine which could be absorbed by the smoker when he smokes the cigarette in standard conditions. The recovery studies were carried out by adding the appropriate volume of standard nicotine solution prepared in supporting electrolyte to the previously determined nicotine content of the tobacco sample. Recovery of nicotine was calculated by comparing the concentration obtained from the spiked mixtures with those of the pure nicotine. It was found that nicotine amounts can be quantitatively recovered by the proposed method, being thus a guarantee of the accuracy of the voltammetric determination of nicotine in tobacco.

Table 3 shows the comparisons of the results for the determination of NIC obtained by several methods [22–26]. Compared with several current techniques for the assay of NIC, CNMCP sensor has some advantages: (1) it has the same specific selectivity but with the advantage of stability and cheap reagents, (2) the

Table 2 The recovery analysis of nicotine in cigarette tobacco

Cigarette brand	[NIC] taken $\times 10^{-6}$ M	[Standard] added $\times 10^{-6}$ M	Found $\times 10^{-6}$ M	Recovery (%)
L&M	60.00	60.00	120.26	100.22
	120.00	–	179.83	99.91
	180.00	–	239.69	99.87
	240.00	–	300.13	100.04
Marlboro	60.00	–	120.30	100.25
	120.00	–	180.06	100.03
	180.00	–	239.98	99.99
	240.00	–	300.19	100.06

Table 3 Comparison of the proposed method with other methods for the determination of NIC

Method	Calibration range (M)	Detection limit (M)	References
HPLC	6.8×10^{-6}–3.4×10^{-5}	6.2×10^{-7}	[22]
SPE	up to 1.2×10^{-6}	1.2×10^{-8}	[23]
CE	8.1×10^{-6}–8.1×10^{-5}	3.8×10^{-7}	[24]
Spectrometry	up to 7.4×10^{-5}		[25]
Flow injection	0–5.8×10^{-2}	6.2×10^{-7}	[26]
CNMCP sensor	$4.0 \times 10^{-6} - 5.0 \times 10^{-4}$	9.4×10^{-8}	This work

instruments of the CNMCP sensor are simpler and cheaper than that of HPLC, GC, SPE and spectrometry, (3) the procedures and pre-treatment for the CNMCP sensor are simple and easy. The results are fairly satisfactory and the background causes no significant effects.

4 Conclusions

In the present work, a biosensor based on CP electrode modified with cerium nanoparticles was used for electrochemical determination of NIC. The advantages of the cerium nanoparticles enhanced the sensitivity of the CP electrode significantly. The experimental conditions such as pH, scan rate, accumulation time and the deposition time were studied to find the highest sensitivity for the determination of NIC. Under the optimum conditions, calibration plots for NIC were linear in the ranges of 4.0×10^{-6}–5.0×10^{-4} M with correlation coefficients of 0.999 and detection limits of 9.43×10^{-8} M. The results showed that the method was simple and sensitive enough for the determination of NIC in cigarettes and urine samples with good precision and accuracy.

References

1. A. Levent, Y. Yardim, Z. Senturk, Voltammetric behavior of nicotine at pencil graphite electrode and its enhancement determination in the presence of anionic surfactant. Electrochim. Acta **55**, 190 (2009)
2. D.J. Doolittle, R. Winegar, J.K. Lee, V. Caldwell, V. Wallace, V.X. Hayes, V. deBethizy, The genotoxic potential of nicotine and its major metabolites. Mutat. Res. **344**, 95 (1995)
3. M. Chena, M. Alegrea, A. Durgavanshib, D. Bosec, J. Romero, Rapid and sensitive determination of nicotine in formulations and biological fluid using micellar liquid chromatography with electrochemical detection. J. Chromatogr. B **878**, 23979 (2010)
4. J.M. Garrigues, A. Perez-Ponce, S. Garrigues, M. de la Guardia, Anal. Fourier-transform infrared determination of nicotine in tobacco samples by transmittance measurements after leaching with CHCl3. Chim. Acta **373**, 63 (1998)
5. N.F. Atta, A. Galal, M. Abu-Attia, S.M. Azab, Carbon paste gold nanoparticles sensor for the selective determination of dopamine in buffered solutions. Journal of ECS **157**(9), F116 (2010)
6. Z. Xie, Q. Liua, Z. Changa, X. Zhang, The developments and challenges of cerium half-cell in zinc–cerium redox flow battery for energy storage. Electrochim. Acta **90**, 695 (2013)
7. S. Liu, J. Li, S. Zhang, J. Zhao, Study on the adsorptive stripping voltammetric determination of trace cerium at a carbon paste electrode modified in situ with cetyltrimethylammonium bromide. Appl. Sur. Sci **252**, 2078 (2005)
8. N.F. Atta, A. Galal, M. Abu-Attia, S.M. Azab, Characterization and electrochemical investigations of micellar/drug interactions. Electrochim. Acta **56**, 2510 (2011)
9. N.K. Chaki, K. Vijayamohanan, Self-assembled monolayers as a tunable platform for biosensor applications. Biosens. Bioelectron. **17**, 1 (2002)
10. Q. Wang, Anodic electrochemical synthesis and characterization of nanocrystalline cerium oxide and cerium oxide/montmorillonite nanocomposites. Dissertation, university of north Texas (2003)

11. Z. Xie, Q. Liu, Z. Chang, X. Zhang, The developments and challenges of cerium half-cell in zinc–cerium redox flow battery for energy storage. Electrochim. Acta **90**, 695 (2013)
12. S.S. Yang, I. Smetana, Evaluation of capillary electrophoresis for the analysis of nicotine and selected minor alkaloids from tobacco. Chromatographia **40**, 375 (1995)
13. X.L. Wen, Y.H. Jia, Z.L. Liu, Micellar effects on the electrochemistry of dopamine and its selective detection in the presence of ascorbic acid. Talanta **50**, 1027 (1999)
14. H.B. Suffredini, M.C. Santos, D. De Souza, L. Codognoto, P. Homem-de-Mello, K.M. Honorio, A.B.F. Da Silva, V. Machado, V. Avaca, Electrochemical behavior of nicotine studied by voltammetric techniques at boron-doped diamond electrodes. Anal. Lett. **38**, 1587 (2005)
15. N. Yang, Q. Wan, J. Yu, Adsorptive voltammetry of Hg(II) ions at a glassy carbon electrode coated with electropolymerized methyl-red film. Sens. Actuators, B **110**, 246 (2005)
16. W. Zhang, G. Xie, S. Li, L. Lu, B. Liu, Au/CeO2–chitosan composite film for hydrogen peroxide sensing. Appl. Surf. Sci. **258**, 8222 (2012)
17. A.M. Fekry, R.R. Mohamed, Acetyl thiourea chitosan as an eco-friendly inhibitor for mild steel in sulphuric acid medium. Electrochim. Acta **55**, 1933 (2010)
18. A.M. Fekry, M.A. Ameer, Electrochemical investigation on corrosion and hydrogen evolution rate of mild steel in sulphuric acid solution. Inter. J. Hydrog. Energy **36**, 11207 (2011)
19. U. Retter, A. Widmann, K. Siegler, H. Kahlert, On the impedance of potassium nickel(II) hexacyanoferrate(II) composite electrodes—the generalization of the Randles model referring to inhomogeneous electrode materials. J. Electroanal. Chem. **546**, 87 (2003)
20. M.A. Ameer, Electrochemical performance of vanadium in sodium chloride solution. Mat. and Corros **51**, 242 (2000)
21. R. Vedalakshmi, V. Saraswathy, S. Ha-Won, N. Palaniswamy, Determination of diffusion coefficient of chloride in concrete using Warburg diffusion coefficient. Corros. Sci. **51**, 1299 (2009)
22. A.Z. Gudet, B. Buri, A simple and rapid HPLC method to quantify and evaluate the purity of nicotine in solution and in microspheres. Pharm. Sci **8**(2), 139 (1998)
23. L.B. Baskin, R.W. Anderson, J.R. Charlson, R.D. Hurt, G.M. Lawson, A solid phase extraction method for determination of nicotine in serum and urine by isotope dilution gas chromatography-mass spectrometry with selected-ion monitoring. Ann. Clin. Biochem. **35**, 522 (1998)
24. G.H. Lu, S. Ralapati, Application of high-performance capillary electrophoresis to the quantitative analysis of nicotine and profiling of other alkaloids in ATF-regulated tobacco products. Electrophoresis **19**, 19 (1997)
25. M.T. Xu, R.F. Qi, Q.M. Gao, Photometric determination of nicotine in tobacco with bromcresal green. Lihua Jianyan, Huaxue Fence **34**, 78 (1998)
26. J.M. Garrigues, A. Perez-Ponce, S. Garrigues, M. de la Guardia, Flow-injection Fourier transform infrared determination of nicotine in tobacco. Analyst **124**, 783 (1999)

Exploring the Antibiotic Effects in Bacterial Biofilms by Epifluorescence and Scanning Electron Microscopy

Luciana Calheiros Gomes, Laura Nunes Silva, Manuel Simões, Luís Ferreira de Melo and Filipe José Mergulhão

Abstract Biofilm bacteria are more resistant to antibiotics than their free-living counterparts. For development of effective antimicrobial therapies, it is crucial to understand the interactions between antibiotics and biofilm cells. This work evaluates the killing and removal effects of ciprofloxacin in *Escherichia coli* biofilms using a combination of epifluorescence and scanning electron microscopy (SEM). When ciprofloxacin was applied to 24 h-old biofilms, two distinct phases were observed. In phase 1 (corresponding to the initial 3 h after application), only 4 % of sessile cells were killed and no removal from the surface was observed. Phase 2 comprises two stages, in the first (between 3 and 4.5 h of antibiotic exposure) about 70 % of the bacterial cells were removed but 60 % of the remaining cells were still alive. In the second stage, it was possible to attain complete cell inactivation (after 7.5 h) although 10 % of the initial cells remained attached to the surface. SEM micrographs show that antibiotic-treated cells have filamentous forms when compared to untreated cells.

L.C. Gomes (✉) · L.N. Silva · M. Simões · L.F. de Melo · F.J. Mergulhão
LEPABE, Department of Chemical Engineering, Faculty of Engineering, University of Porto, Rua Dr. Roberto Frias, S/N, 4200-465 Porto, Portugal
e-mail: luciana.gomes@fe.up.pt

L.N. Silva
e-mail: lauransilva@gmail.com

M. Simões
e-mail: mvs@fe.up.pt

L.F. de Melo
e-mail: lmelo@fe.up.pt

F.J. Mergulhão
e-mail: filipem@fe.up.pt

E.K. Polychroniadis et al. (eds.), *2nd International Multidisciplinary Microscopy and Microanalysis Congress*, Springer Proceedings in Physics 164,
DOI 10.1007/978-3-319-16919-4_31

1 Introduction

Bacteria and other microorganisms tend to attach to living or abiotic surfaces where they grow and produce extracellular polymeric substances, forming a biofilm [1]. Biofilms are responsible for over 60 % of infectious conditions in developed countries where medical implants are frequently used [2]. It is well-known that growth of bacteria in biofilms gives them effective protection not only against the host's defense system but also makes them more resistant to antibiotic treatments [3]. In fact, the minimum inhibitory concentration (MIC) and minimum bactericidal concentration (MBC) of antibiotics to biofilm-growing bacteria may be up to 100–1000 fold higher compared with planktonic bacteria [3]. Biofilm increased tolerance to antibiotics is a complex multifactorial phenomenon where physical, physiological, adaptive and genetic mechanisms play major roles. The three-dimensional structure of the biofilm limits antibiotic diffusion through the matrix and at the same time it can establish a gradient of nutrients and oxygen that originate different physiological states that may affect the antibiotic efficacy [4]. Induction of non-specific or biofilm-specific antibiotic resistance mechanisms, together with the selection of resistance mutations, enhanced by the stress-induced-mutagenesis of the biofilm environment, and horizontal gene transfer, also contribute to biofilm antibiotic resistance [3, 5].

Successful treatment of device-related infections depends on biofilm removal as well as effective killing of sessile bacteria. Traditionally, antibiotics have been used to kill bacteria, although some of them also have the ability to remove cells from surfaces [5–7]. This work assesses the killing and removal effects of ciprofloxacin in *E. coli* biofilms by microscopy with the aim of selecting an appropriate control strategy. Pathogenic strains of *E. coli* are responsible for 70–95 % of urinary tract infections (UTIs), one of the most common bacterial diseases [8]. These infections are usually caused by biofilm development on indwelling urinary catheters and ciprofloxacin is one of the prescribed antibiotics to treat UTIs.

Two different types of microscopic techniques were used in this biofilm study: epifluorescence microscopy and scanning electron microscopy (SEM). Epifluorescence microscopy is used to enumerate the total number of cells when the biofilm is stained with 4′,6-diamidino-2-phenylindole (DAPI), having also the possibility of distinguishing between viable and nonviable cells in a biofilm stained with the Live/Dead® (L/D) *Bac*Light™ kit. DAPI can penetrate an intact cell membrane and bind strongly to A-T rich regions in DNA [9], thus being used to count total sessile cells [10]. The Live/Dead® *Bac*Light™ kit is composed of two nucleic acid-binding dyes, SYTO 9™ and propidium iodide (PI). SYTO 9™ penetrates all bacterial membranes and stains the cells green, while PI only penetrates cells with damaged membranes, and the combination of both stains produces red fluorescing cells. Therefore, bacteria with intact cell membranes stain green and those with damaged membranes stain red [11]. This is a fast epifluorescence staining method to determine the antimicrobial efficacy of antibiotics and biocides against biofilms [5, 12, 13]. SEM was the microscopic technique used to assess the morphological changes in biofilm cells resulting from the antibiotic treatment. Unlike SEM, epifluorescence staining does not allow accurate measurement of the size of bacterial cells since it can only stain specific parts of cells [14]. As a consequence of its excellent resolution properties and wide range of magnifications, SEM yields a three-dimensional rendering of the biofilm surface, revealing the overall shape of the organisms composing the biofilm as well as their organization relative to each other and to the extracellular matrix [15–17].

2 Biofilm Removal and Susceptibility

A volume of 2 mL of *E. coli* JM109(DE3) suspension (1×10^7 cells mL^{-1} in Mueller-Hinton broth) was transferred into each well of a 12-well polystyrene, flat-bottomed microtiter plate containing glass coupons (1 × 1 cm). Initial adhesion and biofilm growth were allowed to occur by incubating the microtiter plates at 37 °C without shaking. At different sampling times, 0.5, 1 and 2 h for initial adhesion and 4, 8, 12, 16, 20 and 24 h for biofilm studies, coupons were removed from the microwells and vortexed in saline solution to resuspend and homogenize the biofilm cells. For total cell counts, suspended cells were filtered through a 25-mm black polycarbonate membrane (pore size 0.2 μm) and stained with 1 mL of DAPI reagent (0.5 mg L^{-1}) for 10 min in the dark [9]. Cells were visualized and counted under an epifluorescence microscope (Leica DM LB2 with a mercury lamp HBO/100 W/3). The microscope incorporates a Leica DFC300 FX camera which acquires images using IM50 software (Leica Microsystems Ltd., Switzerland), a 100× oil immersion fluorescence objective and a filter sensitive to DAPI fluorescence (359-nm excitation filter in combination with a 461-nm emission filter).

To assess the removal and killing effects of antibiotic in biofilms, a concentration corresponding to 5× biofilm MIC of ciprofloxacin (0.4 μg mL^{-1}) was added to the coupons after 24 h of biofilm growth. The biofilm cells were exposed to ciprofloxacin for 7.5 h and different coupons were sampled every 1.5 h. The biofilms were resuspended and stained with the Live/Dead® *Bac*Light™ Bacterial Viability kit (Invitrogen Life Technologies, Alfagene, Portugal) for 10 min in the dark [13]. The optical filter combination consisted of a 480–500-nm excitation filter in combination with a 485-nm emission filter. Both the total number of cells (green plus red cells) and the number of green cells were estimated.

Figure 1 shows the biofilm growth curve and the time-removal and -kill plots for ciprofloxacin against 24-h biofilms. The amount of adhered cells increased linearly over time and the *E. coli* biofilm reached a cell density of about 3×10^7 cm^{-2} after 24 h of incubation. Ciprofloxacin was then applied directly to the biofilm in the following 7.5 h and two distinct effects of antibiotic were detected by epifluorescence microscopy: removal and killing of sessile bacteria. The antibiotic treatment comprises two different phases. In phase 1 (corresponding to the initial 3 h after ciprofloxacin application), a small percentage of microbial cells (4 %) was killed but no removal from the glass surface was observed. Phase 2 includes two stages, in the

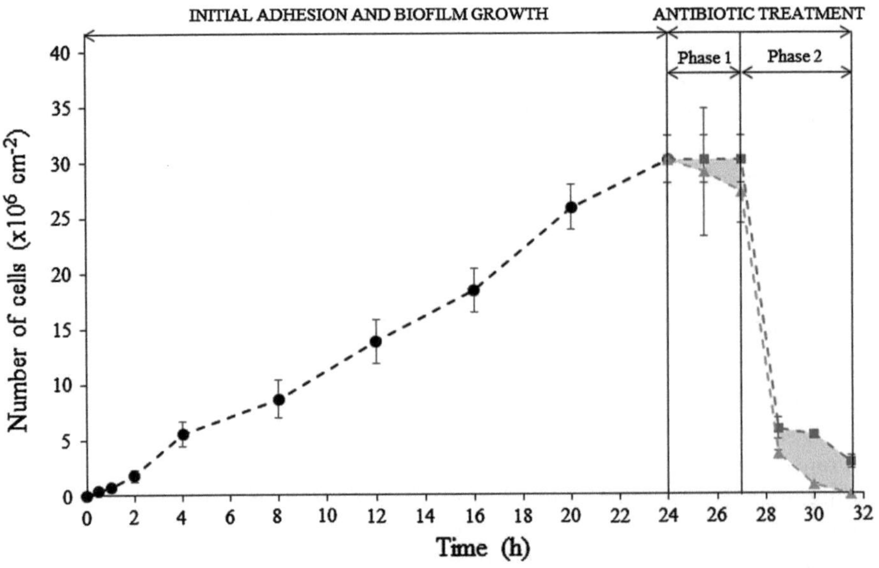

Fig. 1 Time-course evolution of *E. coli* adhesion and biofilm development (*filled circle*), and cell removal (*filled square*) and viability (*filled triangle*) of 24-h biofilms after exposure to 5× biofilm MIC of ciprofloxacin. The shaded regions represent the nonviable cells that remain attached to the surface during the antibiotic treatment. The means ± SDs for three independent experiments are illustrated

first (between 3 and 4.5 h of antibiotic exposure) about 70 % of bacterial cells were removed, however 60 % of the remaining cells were still viable. In the second stage (between 4.5 and 7.5 h of treatment), it was possible to attain complete cell inactivation, although 10 % of the initial cells remained attached to the surface.

These results show that the treatment time should be considered when fighting in vitro biofilms with antibiotics. Traditionally, the biofilm MIC and MBC are the biofilm-related susceptibility parameters used for the selection of antimicrobial agents to treat biofilms formed in clinical settings. However, these criteria only provide a snapshot of the antibiotic effects on the basis of a single concentration and do not consider the antibiotic activity over time after patient administration. Killing-curve approaches provide more appropriate information about the interaction between bacteria and antimicrobial agents since they depict the interaction in a multidimensional way via a dynamic integration of the antibiotic concentration and time of action [18–20]. In contrast with the MIC and MBC determinations, time-kill studies elucidate both the rate and magnitude of bacterial killing. From them, we can obtain key information to establish an efficient dosing regimen and to minimize the development of antimicrobial tolerance and resistance in biofilm infections.

3 SEM Analysis of Biofilm Morphology

The morphological changes of *E. coli* biofilms formed on glass coupons and exposed to 6 h of ciprofloxacin treatment were detected by SEM. Biofilms were fixed in 3 % (w/w) glutaraldehyde in cacodylate buffer, dehydrated with ethanol and hexamethyldisilazane [16], and sputter-coated with a palladium-gold thin film using the SPI Module Sputter Coater equipment for 120 s at 15 mA current. The biofilms were observed with a SEM/EDS system (FEI Quanta 400FEG ESEM/EDAX Genesis X4 M, FEI Company, USA) in high-vacuum mode at 15 kV. Cell length was determined using the microscope software (xT Microscope Control, FEI Company, USA) by measuring 100 randomly selected cells in each condition [16]. Values of cell lenghts were used to obtain a normal size-distribution by best fit parameterization.

The micrographs presented in Fig. 2 confirmed that the ciprofloxacin treatment significantly reduced the total number of initially attached cells, as concluded from the results of epifluorescence microscopy (Fig. 1). The SEM images also revealed that antibiotic-treated cells (Fig. 2(b) and (d)) have filamentous forms when

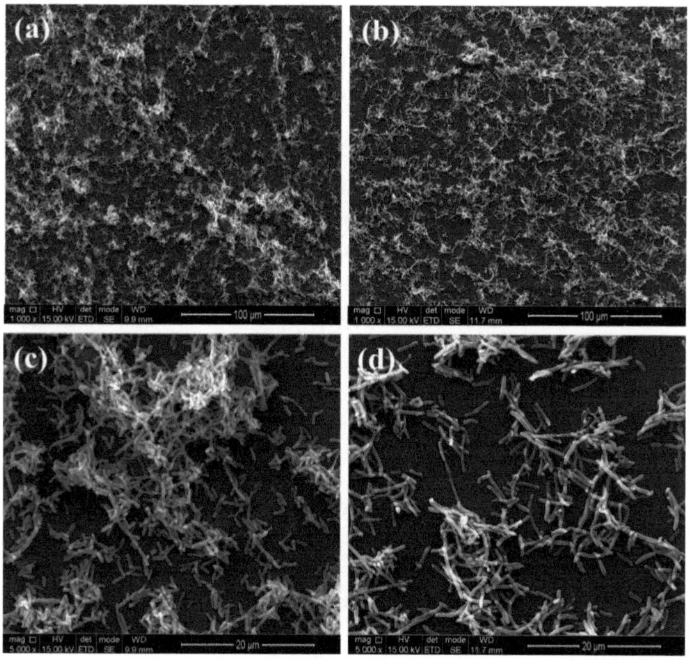

Fig. 2 Scanning electron micrographs of 24-h biofilms not exposed to antibiotic (**a** and **c**) and after 6 h of treatment with 5× biofilm MIC of ciprofloxacin (**b** and **d**). Micrographs (**c**) and (**d**) are high-magnification images (magnification: 5000×; *bars* = 20 μm) of (**a**) and (**b**) (magnification: 1000×; *bars* = 100 μm), respectively

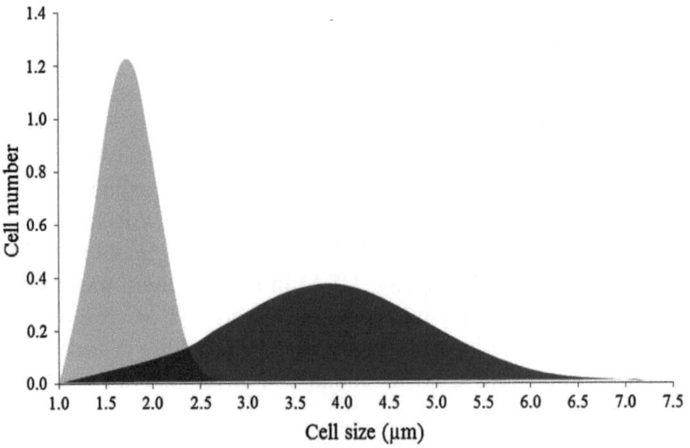

Fig. 3 Cell size distribution of 24-h biofilms not exposed to antibiotic (*light grey shadow*) and after 6 h of treatment with 5× biofilm MIC of ciprofloxacin (*dark grey shadow*)

compared with the control cells (not subjected to the treatment (Fig. 2(a) and (c)). Determination of the cell length in the higher magnification images (Fig. 2(c) and (d)) resulted in a normal curve showing the size distribution of biofilm cells exposed and not exposed to ciprofloxacin (Fig. 3). Although non-exposed cells had lengths ranging from 1.1 to 2.7 μm (on average 1.8 ± 0.3 μm), those exposed to ciprofloxacin measured between 2.0 and 7.1 μm (average cell length of 3.9 ± 1.1 μm). It is well documented that antibiotics can affect bacteria in ways other than the expected bactericidal or bacteriostatic action, in particular they can induce morphological alterations [21–24]. Especially for the exposure to ciprofloxacin, some authors reported the filamentation of *E. coli* cells [25–27] and this phenomenon was associated with the induction of SOS response [23].

4 Conclusions

It is important to evaluate the interactions of bacterial biofilms with antibiotics under conditions mimicking those that exist in clinical settings in order to obtain the most relevant information regarding the efficacy of antimicrobial agents for eradicating biofilm bacteria. This work shows that combining two different microscopy techniques, namely epifluorescence and scanning electron microscopy, can be very useful for the development of efficient antibiofilm treatments.

Acknowledgments The authors acknowledge the financial support provided by Operational Programme for Competitiveness Factors—COMPETE, European Fund for Regional Development—FEDER and by the Portuguese Foundation for Science and Technology—FCT, through Projects PTDC/EBB-BIO/102863/2008 and PTDC/EBB-EBI/105085/2008. Luciana Gomes acknowledges the receipt of a Ph.D. grant from FCT (SFRH/BD/80400/2011).

References

1. P. Stoodley, K. Sauer, D.G. Davies, J.W. Costerton, Biofilms as complex differentiated communities. Annu. Rev. Microbiol. **56**(1), 187–209 (2002)
2. L. Chen, Y. Wen, The role of bacterial biofilm in persistent infections and control strategies. Int. J. Oral Sci. **3**(2), 66–73 (2011)
3. N. Høiby, T. Bjarnsholt, M. Givskov, S. Molin, O. Ciofu, Antibiotic resistance of bacterial biofilms. Int. J. Antimicrob. Agents **35**(4), 322–332 (2010)
4. T.F. Mah, G.A. O'Toole, Mechanisms of biofilm resistance to antimicrobial agents. Trends Microbiol. **9**(1), 34–39 (2001)
5. A. Ito, A. Taniuchi, T. May, K. Kawata, S. Okabe, Increased antibiotic resistance of *Escherichia coli* in mature biofilms. Appl. Environ. Microbiol. **75**(12), 4093–4100 (2009)
6. P.A. Araújo, F.J. Mergulhão, L.F. Melo, M. Simões, The ability of an antimicrobial agent to penetrate a biofilm is not correlated with its killing or removal efficiency. Biofouling **30**(6), 675–683 (2014)
7. B. Passerini de Rossi, C. García, M. Calenda, C. Vay, M. Franco, Activity of levofloxacin and ciprofloxacin on biofilms and planktonic cells of *Stenotrophomonas maltophilia* isolates from patients with device-associated infections. Int. J. Antimicrob. Agents **34**(3), 260–264 (2009)
8. S.M. Jacobsen, D.J. Stickler, H.L.T. Mobley, M.E. Shirtliff, Complicated catheter-associated urinary tract infections due to *Escherichia coli* and *Proteus mirabilis*. Clin. Microbiol. Rev. **21** (1), 26–59 (2008)
9. S. Saby, I. Sibille, L. Mathieu, J.L. Paquin, J.C. Block, Influence of water chlorination on the counting of bacteria with DAPI (4′,6-diamidino-2-phenylindole). Appl. Environ. Microbiol. **63**(4), 1564–1569 (1997)
10. M. Lemos, A. Borges, J.S. Teodósio, P.A. Araújo, F.J. Mergulhão, L.F. Melo, M. Simões, The effects of ferulic and salicylic acids on *Bacillus cereus* and *Pseudomonas fluorescens* single- and dual-species biofilms. Int. Biodeterm. Biodegr. **86, Part A**, 42–51 (2014)
11. Molecular Probes. LIVE/DEAD® *Bac*Light™ Bacterial Viability Kits. 2004
12. M. Simões, M.O. Pereira, M.J. Vieira, Validation of respirometry as a short-term method to assess the efficacy of biocides. Biofouling **21**(1), 9–17 (2005)
13. J.S. Teodósio, M. Simões, F.J. Mergulhão, The influence of nonconjugative *Escherichia coli* plasmids on biofilm formation and resistance. J. Appl. Microbiol. **113**(2), 373–382 (2012)
14. S. Dobretsov, R.M.M. Abed, K. Sharp, O. Skalli, L.G. Boykins, L. Coons, Microscopy of biofilms, in *Biofouling Methods*, ed. by S. Dobretsov, J.C. Thomason, D.N. Williams (Wiley, Oxford, 2014), pp. 1–43
15. M. Alhede, K. Qvortrup, R. Liebrechts, N. Høiby, M. Givskov, T. Bjarnsholt, Combination of microscopic techniques reveals a comprehensive visual impression of biofilm structure and composition. FEMS Immunol. Med. Microbiol. **65**(2), 335–342 (2012)
16. L.C. Gomes, J.M.R. Moreira, J.M. Miranda, M. Simões, L.F. Melo, F.J. Mergulhão, Macroscale *versus* microscale methods for physiological analysis of biofilms formed in 96-well microtiter plates. J. Microbiol. Methods **95**(3), 342–349 (2013)
17. C. Hannig, M. Follo, E. Hellwig, A. Al-Ahmad, Visualization of adherent micro-organisms using different techniques. J. Med. Microbiol. **59**(1), 1–7 (2010)
18. G.A. Pankuch, M.R. Jacobs, P.C. Appelbaum, Study of comparative antipneumococcal activities of penicillin G, RP 59500, erythromycin, sparfloxacin, ciprofloxacin, and

vancomycin by using time-kill methodology. Antimicrob. Agents Chemother. **38**(9), 2065–2072 (1994)
19. M. Mueller, A. de la Peña, H. Derendorf, Issues in pharmacokinetics and pharmacodynamics of anti-infective agents: kill curves versus MIC. Antimicrob. Agents Chemother. **48**(2), 369–377 (2004)
20. E.I. Nielsen, A. Viberg, E. Löwdin, O. Cars, M.O. Karlsson, M. Sandström, Semimechanistic pharmacokinetic/pharmacodynamic model for assessment of activity of antibacterial agents from time-kill curve experiments. Antimicrob. Agents Chemother. **51**(1), 128–136 (2007)
21. O.V. Martinez, H.G. Gratzner, T.I. Malinin, M. Ingram, The effect of some β-lactam antibiotics on *Escherichia coli* studied by flow cytometry. Cytometry **3**(2), 129–133 (1982)
22. M. Jacques, A. Lebrun, B. Foiry, M. Dargis, F. Malouin, Effects of antibiotics on the growth and morphology of *Pasteurella multocida*. J. Gen. Microbiol. **137**(11), 2663–2668 (1991)
23. H.J. Wickens, R.J. Pinney, D.J. Mason, V.A. Gant, Flow cytometric investigation of filamentation, membrane patency, and membrane potential in *Escherichia coli* following ciprofloxacin exposure. Antimicrob. Agents Chemother. **44**(3), 682–687 (2000)
24. Z. Yao, D. Kahne, R. Kishony, Distinct single-cell morphological dynamics under beta-lactam antibiotics. Mol. Cell **48**(5), 705–712 (2012)
25. F. Silva, O. Lourenço, J.A. Queiroz, F.C. Domingues, Bacteriostatic versus bactericidal activity of ciprofloxacin in *Escherichia coli* assessed by flow cytometry using a novel far-red dye. J. Antibiot. **64**(4), 321–325 (2011)
26. D.J. Mason, E.G. Power, H. Talsania, I. Phillips, V.A. Gant, Antibacterial action of ciprofloxacin. Antimicrob. Agents Chemother. **39**(12), 2752–2758 (1995)
27. T.S.J. Elliott, A. Shelton, D. Greenwood, The response of *Escherichia coli* to ciprofloxacin and norfloxacin. J. Med. Microbiol. **23**(1), 83–88 (1987)

Quantitative Confocal Microscopy Analysis as a Basis for Search and Study of Potassium Kv1.x Channel Blockers

Alexey V. Feofanov, Kseniya S. Kudryashova, Oksana V. Nekrasova, Alexander A. Vassilevski, Alexey I. Kuzmenkov, Yuliya V. Korolkova, Eugene V. Grishin and Mikhail P. Kirpichnikov

Abstract Artificial KcsA-Kv1.x (x = 1, 3) receptors were recently designed by transferring the ligand-binding site from human Kv1.x voltage-gated potassium channels into corresponding domain of the bacterial KscA channel. We found that KcsA-Kv1.x receptors expressed in *E. coli* cells are embedded into cell membrane and bind ligands when the cells are transformed to spheroplasts. We supposed that *E. coli* spheroplasts with membrane-embedded KcsA-Kv1.x and fluorescently labeled ligand agitoxin-2 (R-AgTx2) can be used as elements of an advanced analytical system for search and study of Kv1-channel blockers. To realize this idea, special procedures were developed for measurement and quantitative treatment of fluorescence signals obtained from spheroplast membrane using confocal laser scanning microscopy (CLSM). The worked out analytical "mix and read" systems supported by quantitative CLSM analysis were demonstrated to be reliable alternative to radioligand and electrophysiology techniques in the search and study of selective Kv1.x channel blockers of high scientific and medical importance.

1 Introduction

Human voltage-gated potassium Kv1.x-channels (x = 1–3, 5–8) conduct selectively K^+ ions across cell membranes upon depolarization and play an important role in proper cellular functioning [1]. For example, in T lymphocytes, Kv1.3-channel modulates interferon-γ and interleukin 2 production, calcium signaling and proliferation. Kv1.1-channel, which is widely presented in central nervous system, is

A.V. Feofanov (✉) · K.S. Kudryashova · O.V. Nekrasova · A.A. Vassilevski · A.I. Kuzmenkov · Y.V. Korolkova · E.V. Grishin · M.P. Kirpichnikov
Shemyakin-Ovchinnikov Institute of Bioorganic Chemistry, Russian Academy of Sciences, Ul. Miklukho-Maklaya 16/10, 117997 Moscow, Russia
e-mail: avfeofanov@yandex.ru

A.V. Feofanov · K.S. Kudryashova · M.P. Kirpichnikov
Biological Faculty, Lomonosov Moscow State University, Leninskie Gory 1, 119992 Moscow, Russia

E.K. Polychroniadis et al. (eds.), *2nd International Multidisciplinary Microscopy and Microanalysis Congress*, Springer Proceedings in Physics 164,
DOI 10.1007/978-3-319-16919-4_32

involved in controlling of neuronal excitability. Kv1.3 is a target for the treatment of psoriasis, multiple sclerosis, rheumatoid arthritis and type-1 diabetes. Kv1.1 is a target for the treatment of multiple sclerosis and spinal cord injury. Blockers of Kv1.3 and Kv1.1 reduce abnormal activity of the channels and are considered as potential drugs. Moreover, high affinity blockers are molecular tools for structure-functional studies of Kv1-channels.

Conventional approaches to seeking Kv1.x ligands are electrophysiological (patch-clamp) techniques and radioligand binding analysis. Taking into account that besides trivial imaging a confocal laser scanning microscopy (CLSM) can quantitatively measure fluorescent drugs in cells and tissues [2–6], we have tried to develop a convenient fluorescent system for analysis of ligand interactions with Kv1.x-channels. Here we present new robust and easy-to-use analytical systems for search and study of Kv1-blockers, which are based on a quantitative confocal microscopy analysis.

2 Materials and Methods

Bacterial *E. coli* cells, which synthesize artificial KcsA-Kv1.1 or KcsA-Kv1.3 receptors were cultured as described elsewhere [7, 8]. Preparation of spheroplasts from the *E. coli* cells was performed as published earlier [9]. All experiments with spheroplasts were carried out in the buffer: 50 mM Tris-HCl (pH 7.5), 0.1 % bovine serum albumin, 0.25 M sucrose, 10 mM $MgCl_2$, 4 mM KCl, 50 mM NaCl, 0.3 mM EDTA. General protocols for incubation of spheroplasts with a fluorescently labeled agitoxin-2 and competitive ligands can be found elsewhere [9]. Tetraethylammonium chloride (TEA) and recombinant kaliotoxin (KTX) were obtained from Sigma-Aldrich (USA). Recombinant OSK1, agitoxin-2 (AgTx2) and AgTx2 labeled with tetramethylrhodamine (R-AgTx2) were produced by us and purified as described earlier [9].

CLSM measurements were performed with the LSM 710 microscope (Zeiss, Germany) using the αPlan-Apochromat objective (NA 1.46, ×63) and the 543.5 nm excitation wavelength. Fluorescence of R-AgTx2 was registered within the 550–685 nm range. To perform the quantitative analysis of fluorescence associated with spheroplasts, all parameters of the microscope that could affect intensity of the recorded signal were optimized and fixed.

3 Results and Discussion

3.1 Design of Bioengineering Fluorescent System

Recently it was demonstrated that artificial KcsA-Kv1.x (x = 1, 3) receptors can be created by transferring the ligand-binding site from an eukaryotic Kv1.x voltage-gated potassium channel into the appropriated domain of the bacterial KscA

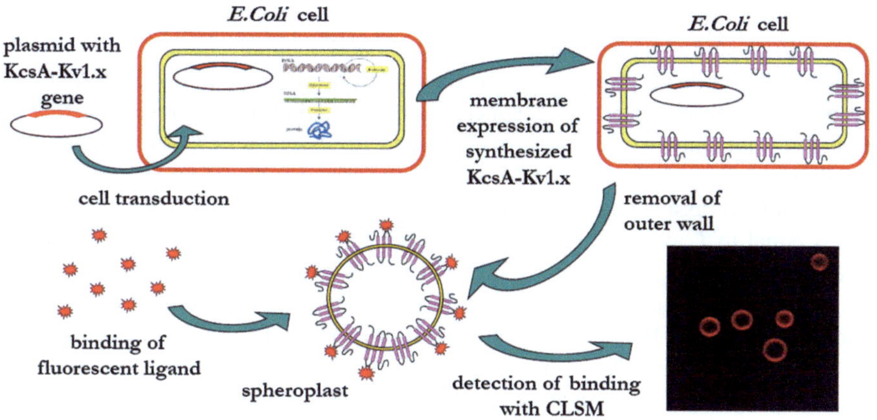

Fig. 1 Scheme of bioengineering fluorescent system. KcsA-Kv1.x (x = 1, 3) proteins expressed in the membrane of *E. coli* expose their ligand-binding sites into periplasmic space and bind ligands when bacterial cell wall is removed. This binding is recorded with CLSM

channel [10, 11]. We found [7] that KcsA-Kv1.x (x = 1, 3), which are synthesized in *E. coli* as membrane-embedded proteins expose their ligand-binding sites in extracellular space and bind external ligands if the outer cell wall is removed (Fig. 1). Accordingly, we have developed new analytical systems that are based on the use of *E. coli* spheroplasts (cells without outer wall) bearing KcsA-Kv1.x membrane receptors, fluorescent ligand R-AgTx2 and CLSM as a method for measuring the amount of complexes formed between R-AgTx2 and KcsA-Kv1.x (Fig. 1).

The advantages of these bioengineering fluorescent systems include short reproduction time of *E. coli*, high growth rates, high levels of protein expression, simple procedures and cheap reagents for cell cultivation; elimination of the time- and money-consuming purification of proteins; physiological environment of receptors without detergents; the mix-and-read regime of measurements; absence of radioactivity.

3.2 A Role of CLSM

Special procedures were developed for measurement and quantitative treatment of fluorescence signals associated with spheroplast membrane using CLSM. CLSM is used [7, 9, 12, 13]: to recognize binding of R-AgTx2 to KcsA-Kv1.x in spheroplast membrane; to control the level of KcsA-Kv1.x expression in cells; to verify integrity of spheroplast membrane; to measure saturated binding curves and determine dissociation constants of complexes between KcsA-Kv1.x and R-AgTx2; to recognize Kv1-channel ligands in complex mixtures and among individual compounds due to displacement of R-AgTx2 from complexes with KcsA-Kv1.x; to measure dissociation constants of examinee ligands using the competitive binding approach; to study characteristic features of ligand-channel binding.

R-AgTx2 binds very specifically to KcsA-Kv1.x, and no binding of R-AgTx2 to membrane components of *E. coli* spheroplasts or to spheroplasts presenting KcsA is observed (not shown). Binding of R-AgTx2 to KcsA-Kv1.x on a spheroplast membrane is recorded with LSCM and characterized with (i) an average fluorescence intensity associated with a particular spheroplast (I_f) and (ii) parameter I_{av} that is equal to I_f averaged over the measured set of spheroplasts.

Dependence of I_{av} on the R-AgTx2 concentration (Fig. 2a) is typical in shape for saturation binding curves that describe ligand-receptor interactions of the 1:1 stoichiometry. Since the frequency distribution of I_f is normal (Gaussian), the spheroplasts are uniform over KcsA-Kv1.x amount, and the I_{av} values are directly proportional to the average quantity of complexes formed by R-AgTx2 with KcsA-Kv1.x.

When the measurement of a saturation binding curve is performed at [L] ≫ [R], where [L] and [R] are concentrations of R-AgTx2 and KcsA-Kv1.x, respectively, this curve can be described with the following equation:

$$I_{av}([L]) = I_{sat}[L]/(K_d + [L]), \tag{1}$$

where I_{sat} is equal to the I_{av} value at the plateau. From this analysis the dissociation constants K_d can be evaluated for the complexes between R-AgTx2 and KcsA-Kv1.x.

Capability of an examinee compound to bind to KcsA-Kv1.x is defined by IC_{50}, which is equal to the concentration of this compound displacing 50 % of R-AgTx2 from complexes with KcsA-Kv1.x. To evaluate IC_{50}, a competitive binding titration is executed by incubating spheroplasts with a constant concentration of R-AgTx2 and increasing concentrations of the examinee compound (Fig. 2b). When the IC_{50} value is obtained from the analysis of CLSM data, the apparent dissociation constant K_i of the examinee compound is calculated with the Cheng-Prusoff equation:

$$K_i = IC_{50}(1 + [L^*]/K_d), \tag{2}$$

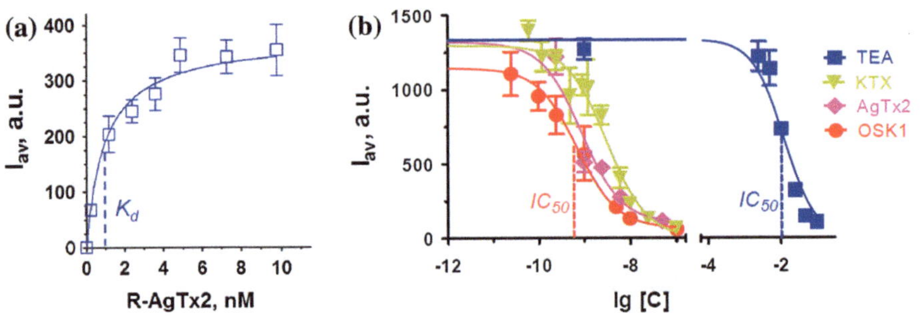

Fig. 2 **a** Typical dependence of I_{av} on the R-AgTx2 concentration. **b** Competition between R-AgTx2 and different ligands for binding to KcsA-Kv1.3 on spheroplasts

where $[L^*]$ is the concentration of free R-AgTx2, which is approximately equal to the total concentration of R-AgTx2 in our conditions.

It was revealed that K_i values determined with CLSM for the complexes of KcsA-Kv1.x (x = 1, 3) with known ligands (both peptides and small organic molecules) are close to the corresponding K_d obtained for these ligands in patch-clamp measurements on Kv1.x (x = 1, 3) channels.

4 Conclusions

These bioengineering systems, which make use of CLSM for data acquisition and processing, are a powerful alternative to radioligand and patch-clamp techniques.

They enable one to search for Kv1.x ligands both among individual compounds and in complex mixtures like animal venoms, as well as to characterize affinities of discovered ligands to Kv1.x channels.

The developed systems will assist in the expansion of Kv1.x ligand library and help to identify selective blockers of Kv1.x channels suitable for drug development.

The principle of a bioengineering system based on presentation of ligand-binding sites at the surface of bacterial membrane can be applied for different eukaryotic proteins, and confocal fluorescence microscopy makes such systems convenient in use.

Acknowledgments This work was supported by Russian Foundation for Basic Research (grant 13-04-01857) and in part by M.V. Lomonosov Moscow State University Program of Development. A.A.V., A.I.K., Y.V.K. and M.P.K. were supported by the Program for fundamental research of the Presidium of Russian Academy of Sciences.

References

1. M.J. Christie, Molecular and functional diversity of K+channels. Clin. Exp. Pharmacol. Physiol. **22**, 944–951 (1995)
2. G.V. Sharonov, T.A. Karmakova, R. Kassies, A.D. Pljutinskaya, M.A. Grin, M. Refregiers, R.I. Yakubovskaya, A.F. Mironov, J.-C. Maurizot, P. Vigny, C. Otto, A.V. Feofanov, Cycloimide bacteriochlorin derivatives: photodynamic properties, cellular- and tissue distribution. Free Radic. Biol. Med. **40**, 407–419 (2006)
3. A. Feofanov, A. Grichine, T. Karmakova, N. Kazachkina, E. Pecherskih, R. Yakubovskaya, E. Luk'yanets, V. Derkacheva, M. Egret-Charlier, P. Vigny, Chelating with metal is not essential for antitumor photodynamic activity of sulfonated phthalocyanines. Photochem. Photobiol. **75**, 527–533 (2002)
4. A. Feofanov, G. Sharonov, A. Grichine, T. Karmakova, A. Pljutinskaya, V. Lebedeva, R. Ruziyev, R. Yakubovskaya, A. Mironov, M. Refregier, J.-C. Maurizot, P. Vigny, Comparative study of photodynamic properties of 13,15-N-cycloimide derivatives of chlorin p6. Photochem. Photobiol. **79**, 172–188 (2004)
5. A. Feofanov, A. Grichine, T. Karmakova, A. Pljutinskaya, V. Lebedeva, A. Filyasova, R. Yakubovskaya, A. Mironov, M. Egret-Charlier, P. Vigny, Near-infrared photosensitizer on

the basis of cycloimide derivative of chlorin p6: 13,15-N-(3'-hydroxypropyl)cycloimide chlorin p6. Photochem. Photobiol. **75**, 633–643 (2002)
6. T. Karmakova, A. Feofanov, A. Pankratov, N. Kazachkina, A. Nazarova, R. Yakubovskaya, V. Lebedeva, R. Ruziyev, A. Mironov, J.-C. Maurizot, P. Vigny, Tissue distribution and in vivo photosensitizing activity of 13,15-[N-(3-hydroxypropyl)] cycloimide chlorin p6 and 13,15-(N-methoxy)cycloimide chlorin p6 methyl ester. J. Photochem. Photobiol. B **82**, 28–36 (2006)
7. O.V. Nekrasova, A.A. Ignatova, A.I. Nazarova, A.V. Feofanov, Y.V. Korolkova, E.F. Boldyreva, A.I. Tagvei, E.V. Grishin, A.S. Arseniev, M.P. Kirpichnikov, Recombinant Kv channels at the membrane of Escherichia coli bind specifically agitoxin2. J. Neuroimmune Pharmacol. **4**, 83–91 (2009)
8. O. Nekrasova, A. Tagway, A. Ignatova, A. Feofanov, M. Kirpichnikov, Studying of membrane localization of recombinant potassium channels in E.coli. Acta Naturae **1**, 91–95 (2009)
9. K.S. Kudryashova, O.V. Nekrasova, A.I. Kuzmenkov, A.A. Vassilevski, A.A. Ignatova, Y.V. Korolkova, E.V. Grishin, M.P. Kirpichnikov, A.V. Feofanov, Fluorescent system based on bacterial expression of hybrid KcsA channels designed for Kv1.3 ligand screening and study. Anal. Bioanal. Chem. **405**, 2379–2389 (2013)
10. C. Legros, V. Pollmann, H.G. Knaus, A.M. Farrell, H. Darbon, P.E. Bougis, M.F. Martin-Eauclaire, O. Pongs, Generating a high affinity scorpion toxin receptor in KcsA-Kv1.3 chimeric potassium channels. J. Biol. Chem. **275**, 16918–16924 (2000)
11. C. Legros, C. Schulze, M.L. Garcia, P.E. Bougis, M.-F. Martin-Eauclaire, O. Pongs, Engineering-specific pharmacological binding sites for peptidyl inhibitors of potassium channels into KcsA. Biochemistry **41**, 15369–15375 (2002)
12. H.N. Anh, V.D.M. Hoang, N.P. Nguyen, K.S. Kudryashova, O.V. Nekrasova, A.V. Feofanov, M.P. Kirpichnikov, T.V. Andreeva, M.V. Serebryakova, V.I. Tsetlin, Y.N. Utkin, Vietnamese heterometrus laoticus scorpion venom: evidence for analgesic and anti-inflammatory activity and isolation of new polypeptide toxin acting on Kv1.3 potassium channel. Toxicon **77**, 40–48 (2014)
13. H.N. Anh, V.D.M. Hoang, K.S. Kudryashova, O.V. Nekrasova, A.V. Feofanov, T.V. Andreeva, V.I. Tsetlin, Y.N. Utkin, Hetlaxin, a new toxin from the heterometrus laoticus scorpion venom, interacts with voltage-gated potassium channel Kv1.3. Dokl. Biochem. Biophys. **449**, 109–111 (2013)

Analysis of Nucleosome Transcription Using Single-Particle FRET

Alexey V. Feofanov, Kseniya S. Kudryashova, Oleg V. Chertkov, Dmitry V. Nikitin, Nikolai A. Pestov, Olga I. Kulaeva, Vasily M. Studitsky and Mikhail P. Kirpichnikov

Abstract Many biological reactions including transcription of a gene are too complex and heterogeneous to be understood by studying ensembles of interacting molecules. In these cases analysis of single complexes can clarify structural and dynamic aspects of these processes. Here we report that single-particle Förster resonance energy transfer (spFRET) microscopy is applicable to investigation of transcription through nucleosomes by an RNA polymerase. Mononucleosomes that support transcription were assembled from core histones and short DNA containing the T7A1 promoter and strong 603 nucleosome-positioning sequence. Fluorophores (Cy3 and Cy5) were introduced in the neighboring coils of nucleosome DNA in spatially close positions without disturbance of nucleosomal structure or transcription. Such labeling allows the changes in the Cy3–Cy5 distance caused by DNA uncoiling from the octamer or DNA looping to be monitored as changes in FRET efficiency. spFRET measurements for freely diffusing single nucleosomes were conducted using a laser scanning confocal microscope equipped with avalanche photodiodes. Nucleosome subpopulations that differ in FRET efficiency (i.e. in nucleosome structure) were revealed. RNA polymerase was stalled in distinct positions on the nucleosomal DNA during transcription, and the structures of these complexes were characterized with spFRET microscopy.

A.V. Feofanov (✉) · K.S. Kudryashova · O.V. Chertkov · D.V. Nikitin · N.A. Pestov · O.I. Kulaeva · V.M. Studitsky · M.P. Kirpichnikov
Biological Faculty, Lomonosov Moscow State University, Leninskie Gory 1, 119992 Moscow, Russia
e-mail: avfeofanov@yandex.ru

V.M. Studitsky
e-mail: Vasily.Studitsky@fccc.edu

A.V. Feofanov · K.S. Kudryashova · O.V. Chertkov · M.P. Kirpichnikov
Shemyakin-Ovchinnikov Institute of Bioorganic Chemistry, Russian Academy of Sciences, ul. Miklukho-Maklaya 16/10, 117997 Moscow, Russia

O.I. Kulaeva · V.M. Studitsky
Cancer Epigenetics Program, Fox Chase Cancer Center, Philadelphia, USA

E.K. Polychroniadis et al. (eds.), *2nd International Multidisciplinary Microscopy and Microanalysis Congress*, Springer Proceedings in Physics 164,
DOI 10.1007/978-3-319-16919-4_33

1 Introduction

Chromatin transcription by RNA polymerase (RNAP) is a complex process that requires tightly regulated DNA structure reorganization in the course of transcription and recovery of the initial structure after its completion [1]. Many important aspects of this process can be unraveled with high spatial and temporal resolution using model mononucleosomes assembled from core histones and short DNA containing promoter (such as T7A1) and strong nucleosome-positioning sequence (such as 603 template [1]). An advanced approach to the study of structural features of transcription in the mononucleosome system can be realized on the basis of single-particle Förster resonance energy transfer (spFRET) microscopy.

Here we report that donor-acceptor pair of fluorophores can be introduced in DNA for the purposes of spFRET analysis without disturbance of transcription by RNAP. We present the results of comparison of structural organization of DNA in mononucleosomes before and after transcription as well as in complexes containing stalled RNAP using spFRET microscopy as a probing tool.

2 Materials and Methods

Oligonucleotides and oligonucleotides labeled with Cy3 and Cy5 fluorophores were obtained from Sintol (Russia). Fluorophore-labeled 603 DNA template was produced in two sequential annealing steps. *3DW_Cy3_603-5A2* and *5DW_Cy5_603-5A2* primers were annealed, and the product was annealed with *6035AFP* and *6035ARP* primers.

3DW_Cy3_603-5A2: GCGTCTAAGTACGC**T(Cy3)**TAGCGCACGGTAGA GC GCAATCCAAGGCTAACCACCGTGCATCGATGTTGAAAGAGGCCCTC CGT CCTTATTACTTCAAGTCCCTGGGGT

5DW_Cy5_603-5A2: CCCCAGGGACT**T(Cy5)**GAAGTAATAAGGACGGA GGGCCTCTTTCAACATCGATGCACGGTGGTTAGCCTTGGATTGCGCTC TACCGTGCGCTAAGCGTA

6035AFP: CCCGGTTCGCGCGCCCGCCTTCCGTGTGTTGTCTCCTCTCGG

6035ARP: CTTAGACGCCCGAGTGACGACTTCACTCGGCAGGCGGGC GCGCGAACCGGGCCCAGTGCC

The final duplex was treated by bacteriophage T4 DNA ligase to avoid any nicks, ligated to T7A1 promoter fragment by TspRI sites and gel-purified.

For reconstitution of nucleosomes the H1-depleted (−H1) donor chromatin was obtained as described elsewhere [2]. Nucleosomes were assembled on the DNA template using −H1 chromatin as a donor of the histone octamer by dialysis against decreasing concentrations of NaCl. RNA-polymerase was purified from *E. coli* cells as described earlier [3].

Open complex (OC) was produced by addition of AUC initiating triribonucleotide (20 μM) to the mixture of the assembled nucleosomes (200 ng in 20 μl of buffer containing 40 mM KCl) with RNAP (fivefold molar excess). Formation of the complex stalled at position −5 (EC-5) relative to a promoter-proximal nucleosome boundary was performed by addition of ATP, GTP and CTP (1 μM) to OC. Complexes were purified with Ni^{2+}–NTA agarose beads as described elsewhere [4]. Formation of OC and EC-5 was verified by the analysis of synthesized radioactively labeled RNA products on denaturing 8 % polyacrylamide gels as described ealier [4]. All four ribonucleotides (4 mM) were added to purified EC-5 to start the chase reaction, and the resulting reaction mixture containing post chase (PC) sample was analyzed. For spFRET measurements specimens were diluted to 0.2 nM with the buffer 20 mM Tris–HCl (pH 8.0), 5 mM $MgCl_2$, 2 mM 2-mercaptoethanol, 150 mM KCl, 0.1 % poly(ethylene glycol) (*ca.* 400 Da).

spFRET measurements were performed with the LSM710-Confocor3 laser scanning confocal microscope (Zeiss, Germany) using the C-Apochromat water-immersion objective (40×, 1.2 NA). Fluorescence was excited with the 514.5 nm laser line of an Ar^{+}-ion laser (2 μW under objective). Diameter of a confocal diaphragm corresponded to 1 Airy disk. Fluorescence of Cy3 and Cy5 was recorded with avalanche photodiodes in the 530–635 and 635–800 nm ranges, respectively.

Time lapses for nucleosome measurements were 10 min (conventional experiments) and 30 min (specimen stability studies) with dwell time of 5 ms. Thresholds for discrimination of nucleosome signals were $10 < I_3 < 80$ kHz or $5 < I_5 < 80$ kHz, where I_3 and I_5 are signal intensities in Cy3 and Cy5 channels, respectively. Background intensities were $I_3 = 1.0 \pm 0.1$ kHz and $I_5 = 0.5 \pm 0.1$ kHz.

The sets of discriminated Cy3 and Cy5 intensities were recalculated into sets of FRET efficiencies (E) using equation

$$E = (I_5 - 0.19 \times I_3)/(I_5 + 0.81 \times I_3), \tag{1}$$

where factors 0.19 and 0.81 provided correction for the contribution of Cy3 fluorescence in the Cy5 detection channel (spectral cross-talk). Samplings of nucleosomes contained at least 6000 particles. Sets of FRET efficiencies calculated for nucleosome samplings were presented as frequency distribution histograms and fitted with a sum of Gaussian bands.

3 Results and Discussion

spFRET measurements were performed in very diluted solutions of nucleosomes (see Materials and methods) where nucleosomes diffused freely through the laser focal volume one by one, and signals from single particles were resolved in time. spFRET data were collected from thousands of nucleosomes. Histograms describing distribution of *E*-values for single nucleosomes or their complexes with RNAP are presented in Fig. 1. The main fraction of fluorescently labeled

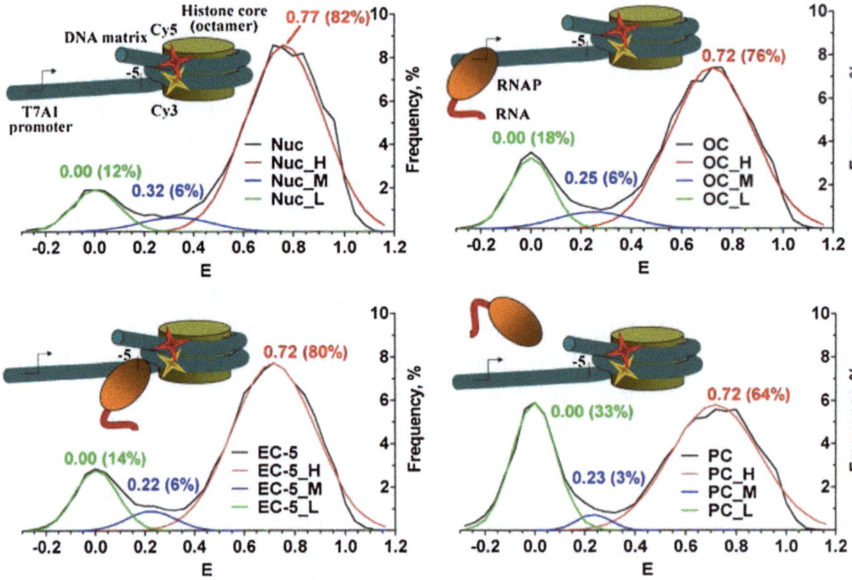

Fig. 1 spFRET analysis of nucleosomes and their stalled complexes with RNAP. Frequency histograms of FRET efficiency (E) are presented for nucleosomes (Nuc), OC, EC-5 and PC. Color bands are the result of histogram deconvolution into three states: *high* (H), *middle* (M) and *low* (L) FRET. *E*-values and relative fractions of nucleosomes (in parentheses) are indicated for each state

nucleosomes has a high *E*-value that indicates that Cy3 and Cy5 fluorophores in the neighboring DNA coils are in close proximity in the compact nucleosome structure. Fluorophore-labeled nucleosomes preserve ability to form complexes with RNAP as confirmed by the analysis of synthesized radioactively labeled RNA products (not shown). Moreover, transcription through the fluorophore-labeled nucleosome occurs, and the distributions of intermediates formed during transcription are very similar for labeled and non-labeled nucleosomes (not shown). Therefore, fluorophores do not interfere with RNAP progression, and fluorophore-labeled nucleosomes can be utilized as a new instrument for investigation of transcription with spFRET microscopy.

spFRET analysis shows that nucleosomes, OC and EC-5 complexes contain three subpopulations of particles with low (L), middle (M) and high (H) FRET (Fig. 1). It should be noted that the existence and size of these subpopulations cannot be revealed with other methods including conventional FRET measurements from a nucleosome ensemble. The main subpopulation of particles with H-FRET is intact nucleosomes with the compact DNA structure in the regions of Cy3 and Cy5 locations. L-FRET subpopulation ($E = 0.00$) consists of unwrapped nucleosomes and free DNA matrix. M-FRET particles were concluded [5] to be hexasomes [6] formed after the H2A/H2B histone dimer dissociation. Distal DNA tail is partially unwrapped in hexasomes that increases the Cy3–Cy5 distance and decreases *E*.

RNAP binding to T7A1 promoter and formation of OC decreases slightly the *E*-value of H-FRET state (Fig. 1 A) that can be explained by a sterical interference between bulk RNAP and a distal DNA end leading to the disturbance of DNA packing and small increase in the Cy3–Cy5 distance. RNAP propagation from OC to EC-5 does not induce additional changes of DNA packing in the distal region of a nucleosome (Fig. 1). These spFRET microscopy data are consistent with the biochemical data indicating that a nucleosome structure is minimally perturbed in the EC-5 complex [1].

PC state which is formed after nucleosome transcription by RNAP consists also of L-, M- and H- FRET subpopulations of nucleosomes (Fig. 1). According to biochemical data >95 % of EC-5 complexes are transcribed (not shown), and spFRET analysis shows that histons dissociate from DNA (at least partially) just in *ca.* 16 % of transcribed nucleosomes (as judged from L-FRET subpopulation increase, Fig. 1). A main fraction of H-FRET nucleosomes (64 %) survives during transcription and preserves compact DNA packing which is similar to that of EC-5.

An attempt to study conformational changes occurring during transcription in freely diffusing nucleosomes is illustrated in Fig. 2. A chase reaction was initiated under a microscope by addition of NTPs (50 μM) to the EC-5 in solution. Delay between NTPs addition and initiation of measurements was *ca.* 3 s. To provide statistical analysis (sampling of 500–600 nucleosomes) it was necessary to collect data from diffusing nucleosomes during at least 10 s periods and repeat 2–3 times the chase reaction with fresh EC-5 complexes. Comparison of histograms for different 10 s periods reveals considerable changes in the shape of *E*-profiles (Fig. 2) but interpretation of these changes in terms of intermediates of transcription is complicated because of two reasons. First, transcription of nucleosome by RNAP occurs quickly (up to 30 nucleotides per second on histone-free DNA), whereas time resolution of this spFRET approach is 10 s. Second, transcription in nucleosomes occurs asynchronously, and shortly after initiation of transcription different nucleosomes are present at different stages of the process. It means that spFRET

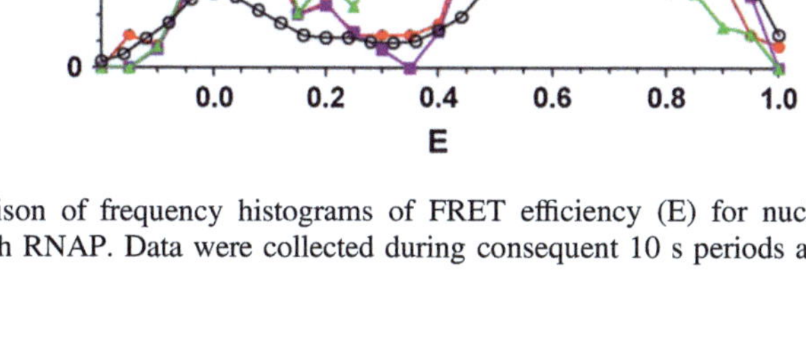

Fig. 2 Comparison of frequency histograms of FRET efficiency (E) for nucleosomes during transcription with RNAP. Data were collected during consequent 10 s periods after transcription initiation

microscopy of freely diffusing nucleosomes is not suitable for time-resolved study of transcription but well applicable for the investigation of stationary (equilibrium) states like stalled complexes and very slow reactions. To study transcription with high temporal resolution (100 ms and less) using spFRET microscopy, the nucleosomes should be immobilized, and these experiments are in progress.

Fluorophore-labeled nucleosomes, their OC and EC-5 are very stable in the buffers containing up to 150 mM KCl: H-FRET subpopulation of nucleosomes is preserved unchanged during at least 30 min of measurements (not shown). This stability is a solid basis for the ongoing detailed spFRET microscopy studies of nucleosome transcription by RNAP.

Acknowledgments This work was supported by Russian Science Foundation (grant 14-24-00031).

References

1. O.I. Kulaeva, D.A. Gaykalova, N.A. Pestov, V.V. Golovastov, D.G. Vassylyev, I. Artsimovitch, V.M. Studitsky, Mechanism of chromatin remodeling and recovery during passage of RNA polymerase II. Nat. Struct. Mol. Biol. **16**, 1272–1278 (2009)
2. D.A. Gaykalova, O.I. Kulaeva, V.A. Bondarenko, V.M. Studitsky, Preparation and analysis of uniquely positioned mononucleosomes. Methods Mol. Biol. **523**, 109–123 (2009)
3. W. Walter, M.L. Kireeva, V. Tchernajenko, M. Kashlev, V.M. Studitsky, Assay of the fate of the nucleosome during transcription by RNA polymerase II. Methods Enzymol. **371**, 564–577 (2003)
4. D.A. Gaykalova, O.I. Kulaeva, N.A. Pestov, F.K. Hsieh, V.M. Studitsky, Experimental analysis of the mechanism of chromatin remodeling by RNA polymerase II. Methods Enzymol. **512**, 293–314 (2012)
5. A. Gansen, A. Valeri, F. Hauger, S. Felekyan, S. Kalinin, K. Toth, J. Langowski, C.A. Seidel, Nucleosome disassembly intermediates characterized by single-molecule FRET. Proc. Natl. Acad. Sci. U.S.A. **106**, 15308–15313 (2009)
6. M.L. Kireeva, W. Walter, V. Tchernajenko, V. Bondarenko, M. Kashlev, V.M. Studitsky, Nucleosome remodeling induced by RNA polymerase II: loss of the H2A/H2B dimer during transcription. Mol. Cell **9**, 541–552 (2002)

Index

A
Amide, 146, 149
ARM200F, 19, 20
Auger spectroscopy, 175, 176
Automation, 58, 63

B
Bio-imaging, 155
Bismuth antimony telluride, 4
Blockers, 250, 253
Boron doping, 177

C
Ca, 218–220
Classification, 62
Cold-FEG, 19–21, 23
Collagen, 218–221
Computer simulations, 153
Conducting Atomic Force Microscopy, 9, 15
Confocal, 250, 253
Confocal fluorescence microscopy, 153
Corrosion resistance, 208, 213
Corrosion, 197–199
Crystalization, 73, 77

D
Diatomite, 73, 74, 79–82, 86, 87
Differential pinhole, 157
Diffusion barrier, 169
Digital holographic microscopy, 92
Dissociation constants, 251, 252

E
EBSD, 176, 179
EDS, 218, 219
EDX, 146–148, 175, 176
Electrochemical technique, 198
ESEM, 146, 147

F
Fe-Al-Mn system, 205, 206
Fluorescence, 257
Fractal analysis, 81

G
Garnet, 65–67
Grain refinement, 176

H
High resolution, 153
High temperature reliability, 169
Holographic particle image velocimetry, 92
Holography, 92

I
Image analysis, 61, 62, 81

J
JEOL, 20, 23

L
^{6}LiF, 162, 163
Liver fibrosis, 223

M
Magnetization, 66
Mechanical properties, 51, 55
Mercury porosimetry, 80, 82
Mesenchymal stem cells, 223–225, 227, 228

N
Non-volatile memory cells, 9
Nucleosome, 255–260

P
Peptides, 253
Performance improvement, 157

E.K. Polychroniadis et al. (eds.), *2nd International Multidisciplinary Microscopy and Microanalysis Congress*, Springer Proceedings in Physics 164,
DOI 10.1007/978-3-319-16919-4

Phase analysis, 177
Pore space, 79–81, 83
Potassium channels, 249

Q
Quality assurance, 58, 62

R
RNA polymerase, 255, 256

S
Salt fibrous, 74
SEM, 52, 175–177
SEM-EDS, 74
Sensitization, 197, 198, 200, 202
Siloxane based scintillators, 162
Sintering, 52, 53, 55
Sol-gel, 66, 69
spFRET, 255–259
Stainless steels, 205, 208

T
γ-TiAl based intermetallics, 175
Tantalum, 170
TbIG, 65–69
Thermal neutron detection, 166
Thermoelectric, 3
Thin film, 3, 66
Titanium-tungsten, 170
Transcription, 256, 258, 259
Transmission Electron Microscopy (TEM), 4, 6, 9

V
Vertebra, 218–221

W
Wavelength Depended Speckle Correlation Analysis (WDSCA), 39, 49

MIX
Papier aus verantwortungsvollen Quellen
Paper from responsible sources
FSC® C105338

If you have any concerns about our products,
you can contact us on
ProductSafety@springernature.com

In case Publisher is established outside the EU,
the EU authorized representative is:
Springer Nature Customer Service Center GmbH
Europaplatz 3, 69115 Heidelberg, Germany

Printed by Libri Plureos GmbH
in Hamburg, Germany